Data Warehousing and Data Mining for Telecommunications

Chapter 7   Screen shots reprinted with permission from Microsoft Corporation, One Microsoft Way, Redmond, WA 98052-6399

Chapter 8   Screen shots of PowerPlay reprinted with permission from Cognos Corporation, One Burlington Business Center, 67 So. Bedford Street, Suite 200W, Burlington, MA 01803-5164, Tel: 1-800-365-3968, http://www.cognos.com

Chapter 9   Screen shots reprinted with permission from SPSS Inc., 444 N. Michigan Avenue, Chicago, IL 60611-3962, Tel: 312-329-2400, Fax: 312-329-3690, http://www.spss.com

Chapter 10   Screen shots of ModelMAX reprinted with permission from Advanced Software Applications. ModelMAX is a registered trademark of Advanced Software Applications, 333 Baldwin Road, Pittsburgh, PA 15205, Tel: 412-429-1003, Fax: 412-429-0709, e-mail: asa@fyi.net

Chapter 11   MapInfo Professional™ screen shots provided by MapInfo Corporation. ©1985–1997 MapInfo Corporation. All rights reserved. MapMarker is a registered trademark and MapInfo Professional and StreetWorks are trademarks of MapInfo Corporation, One Global View, Troy, New York 12180-8399, Tel: 518-285-6000, Fax: 518-285-6060, http://www.mapinfo.com
MapInfo Corporation makes no guarantee or warranty with regard to the accuracy of information supplied herein and accepts no liability for loss or damage incurred as a result of any reliance on the information.

Appendix A   Screen shots reprinted with permission from CNET-France Telecom and SLP InfoWare Inc.

Appendix B   Screen shots reprinted with permission from Seagate Software, 1095 West Pender Street, Vancouver, BC, Canada V6E 2M6, Tel: 604–681–3435, Fax: 604–681–2934, http://www.seagatesoftware.com. Seagate and Seagate Technology are registered trademarks of Seagate Technology, Inc. Seagate Software is a trademark of Seagate Technology, Inc. Seagate Software Information Management Group, Seagate Crystal Reports, Seagate Crystal Info, and Seagate Holos are trademarks or registered trademarks of Seagate Technology, Inc., or one of its subsidiaries. All other trademarks and registered trademarks are property of their respective owners.

For a complete listing of the *Artech House Computer Science library*, turn to the back of this book.

Data Warehousing and Data Mining for Telecommunications

Rob Mattison

Artech House
Boston • London

Library of Congress Cataloging-in-Publication Data
Mattison, Rob.
　Data warehousing and data mining for telecommunications / Rob Mattison.
　　p.　cm.—(Artech Computer Science Library)
　Includes bibliographical references and index.
　ISBN 0-89006-952-2 (alk. paper)
　1. Telecommunications—Management.　2. Database Management.　3. Data mining　I. Title.　II. Series
HE7661.M38　　1997
384'.0285'574—dc21　　　　　　　　　　　　97-22863
　　　　　　　　　　　　　　　　　　　　　　CIP

British Library Cataloguing in Publication Data
Mattison, Rob M.
　Data warehousing and data mining for telecommunications—(Artech Computer Science Library)
　1. Decision support systems　2. Telecommunications　3. Digital communications
　I. Title
　384'.0285

ISBN 0-89006-952-2

Cover and text design by Darrell Judd. Illustrations by Brigitte Kilger-Mattison. Cover images from Seagate Software and SLP InfoWare.

© **1997 ARTECH HOUSE, INC.**
685 Canton Street
Norwood, MA 02062

All rights reserved. Printed and bound in the United States of America. No part of this book may be reproduced or utilized in any form or by any means, electronic or mechanical, including photocopying, recording, or by any information storage and retrieval system, without permission in writing from the publisher.
　All terms mentioned in this book that are known to be trademarks or service marks have been appropriately capitalized. Artech House cannot attest to the accuracy of this information. Use of a term in this book should not be regarded as affecting the validity of any trademark or service mark.

International Standard Book Number: 0-89006-952-2
Library of Congress Catalog Card Number: 97-22863

10 9 8 7 6 5 4 3 2

I would like to dedicate this book to my darling grandchildren, Jonathan Yesulis, Raquel Kuykendall, and Anthony and Nicole Cirrincione, and to all of the other children who will be left to pick up the pieces after we move on.

Contents

Foreword *xiii*

Preface *xvii*

Chapter 1
Everything's up to date in Kansas City 1

1.1 The current industry composition 3
1.2 Why is telecommunications so BIG? 4
1.3 Telecommunications: the major driving economic force of the 21st century 5
1.4 Knowledge management enablement—the biggest factor of all 6
1.5 The ultimate environment 7
    1.5.1 Failed excursions into the new frontiers 7

1.6 Future directions 8
1.7 Telecommunications and technological innovation 9
    1.7.1 Peg counts 9
    1.7.2 Business drives technological innovation 9
1.8 The three strategic options 10
1.9 Customer intimacy—from "network is king" to "customer is king" 10
    1.9.1 Marketing as the driving force 11
1.10 Operational efficiency—being the low-cost provider of choice 11
1.11 Technical proficiency—being the best at what you do 12
1.12 Conclusion 12

Chapter 2
Why warehousing and how to get started 13

2.1 Background of data warehousing 14
    2.1.1 The history of the data warehousing phenomenon 14
    2.1.2 Data warehousing—in a nutshell 16
    2.1.3 What is a data warehouse? 17

2.2 Data mining 18
    2.2.1 Why should one seriously consider using these approaches? 20

2.3 Why are these approaches so exceptionally valuable to telecommunications firms? 20
    2.3.1 Data intensity 20
    2.3.2 Analysis dependency 21
    2.3.3 Competitive climate 21
    2.3.4 Technological change at a very high rate 21
    2.3.5 Historical precedent 22

2.4 Organizing the process 22
    2.4.1 An inventory of the existing computer systems and other technological infrastructure 22
    2.4.2 A roadmap and an approach for how to deploy data warehouses in general 23
    2.4.3 A roadmap for understanding how to diagnose and develop a plan for identifying the best things to put into the warehouse and which data mining tools to use 23

Chapter 3

The knowledge management view of business and warehousing 25

3.1 The knowledge management revolution 26
    3.1.1 Knowledge management principles 27
    3.1.2 The organizational footprint and what it tells us about knowledge transformation processes 29

3.2 Efficiency optimization—optimize the silo or optimize the whole 32
    3.2.1 Which type of warehouse is better, or which is the right one? 34
    3.2.2 The warehouse alternative 37
    3.2.3 A third alternative 37

3.3 The corporate global warehouse model 38
    3.3.1 Developing a truly usable global architecture model 40
    3.3.2 An alternative foundation: the value chain 42
    3.3.3 The key to value chain delivery 45

3.4 Overall strategy for development (one piece at a time, fitting into the overall architecture) 45
    3.4.1 The growing warehouse example 47
    3.4.2 Ownership of knowledge issues 49

Chapter 4

The telecommunications value chain 51

4.1 The knowledge roadmap solution 52

4.2 Steps in the process of deriving a business' value chain 52

4.3 Telecommunications functions and systems 53

4.3.1 Creation (new product development and exploitation) 54
4.3.2 Acquisition (acquiring the "right" to do business) 54
4.3.3 Network infrastructure planning and development (creating the "phone system") 55
4.3.4 Network infrastructure maintenance (maintaining the "phone system") 56
4.3.5 Provisioning (setting up customer services) 57
4.3.6 Activation (activating customer services) 57
4.3.7 Service order processing 58
4.3.8 Billing (tracking service and invoicing the customer) 58
4.3.9 Marketing (identifying prospects/channels, advertising) 59
4.3.10 Customer service (keeping the customer happy) 59
4.3.11 Sales (establishing and maintaining customer relationships) 61
4.3.12 Finance and accounting 61
4.3.13 Credit management 61
4.3.14 Operations (network and business) 62
4.3.15 A comprehensive value chain 62

4.4 Organizational structure and the value chain 63
4.4.1 Typical organizational structure: medium-sized cellular firm 64
4.4.2 Typical organizational structure: large telecommunications firms 66

4.5 Allocating the business units to the value chain and the knowledge management process 66

4.5.1 Aligning the value chain and the organization—large megacorporation 68
4.5.2 Aligning the value chain with the information systems 71
4.5.3 Kingpin systems: the beginning of computer systems alignment 72
4.5.4 Alignment problems and their symptoms 74
4.5.5 Data warehousing as an alternative 76
4.5.6 Data warehousing as a migration path 77
4.5.7 The fully aligned model—a summary 77

Chapter 5

Building the warehouse—one step at a time 79

5.1 Challenges to infrastructure design 80
5.2 The functional components of a warehouse environment 84
5.2.1 Acquisition 85
5.2.2 Storage 87
5.2.3 Access 88
5.2.4 The operational infrastructure 89
5.2.5 The physical infrastructure 89

5.3 The step-by-step, cost-justified approach 89
5.3.1 What is a value proposition? 90
5.3.2 Gathering value propositions 90

5.4 How do you build a warehouse? 92

Chapter 6

Value propositions in telecommunications 95

6.1 Mining tools and value delivery 96
    6.1.1 Operational monitoring and control 96
    6.1.2 Discovery and exploration 97

6.2 Value propositions by functional area 98
    6.2.1 Marketing value propositions (historical/cross-silo/discovery) 99
    6.2.2 Credit value propositions 99
    6.2.3 Customer service value propositions—(real-time and historical/cross-silo/operational monitoring) 100
    6.2.4 Sales value propositions 101
    6.2.5 Network planning value propositions 101
    6.2.6 Network maintenance value propositions 103
    6.2.7 Creation 103
    6.2.8 Activation and provisioning and service order processing 104
    6.2.9 Billing (historical/single-silo/discovery and monitoring) 104
    6.2.10 Operations 104

6.3 Conclusions 105
    6.3.1 Knowledge management approach 105

Chapter 7

Simple sales analysis: an introduction to operational monitoring using Microsoft Query 107

7.1 Operational efficiency—an overview 109
7.2 Sales monitoring and control 110
7.3 A universal problem 111
7.4 Using Microsoft Query and Excel to do sales tracking 111
    7.4.1 The sales database 112
7.5 Managing more complicated needs 115
7.6 Alternative methods of accessing data 115

Chapter 8

Sales and product management: advanced operational monitoring using COGNOS PowerPlay 117

8.1 Monitoring complex business organizations 118
    8.1.1 Determining the different levels at which to report 119
    8.1.2 Preparing the data for use 121
8.2 Exploring sales and product performance 121
8.3 Additional PowerPlay features 124
    8.3.1 Alerts 125
    8.3.2 Schedulers 126
8.4 Summary 127

Chapter 9

Customer intimacy: an introduction using SPSS 129

9.1 An introduction to analytical mining 130
9.2 Statistical analysis—options and objectives 131
9.3 Descriptive approaches 133

9.4 Inferential approaches—regression analysis 137
9.5 Conclusions on statistical analysis 140

Chapter 10
Predicting customer behavior: an introduction to neural networks 143

10.1 Unraveling complex situations 144
10.2 How can a neural network help with marketing? 145
10.3 Step-by-step use of a neural network 145
　10.3.1 *What does the training report tell us? 146*
　10.3.2 *Creating and interpreting the gains table 147*
　10.3.3 *Analyzing the gains chart 149*
　10.3.4 *Making marketing programs as profitable as possible 151*
10.4 Applying the model to prospects 152
10.5 Conclusion on neural networks 152

Chapter 11
Engineering and competitive analysis support: an introduction to geographical systems and MapInfo 155

11.1 An introduction to MapInfo Professional 156
　11.1.1 *MapInfo telecommunications offerings 156*
11.2 Using geographical information to solve telecommunications problems 158
11.3 Cellsite analysis with MapInfo Professional 159
11.4 Market analysis capabilities 162
11.5 Viewing a local market in greater detail 164
11.6 Accessibility to fiber analysis 165
11.7 Working with the underlying database 167
11.8 Conclusion 168

Appendix A
Real world warehousing: France Telecom and STATlab tools 169

Appendix B
The business case for business intelligence 211

Appendix C
SPSS 239

Appendix D
The DecisionWORKS suite from Advanced Software Applications 245

Glossary 251

Selected bibliography 257

About the author **261**

Index **263**

Foreword

OVER THE PAST TEN YEARS the telecommunications industry has undergone an incredible growth and development spurt. This growth has affected every line of business in telecommunications, including the long distance business, the emerging wireless industry, and the changing wireline industry. Competition has entered every one of these businesses as expansion has shifted into high gear. And competition is a key attribute of these businesses in the future. Who will survive and even what the future telecommunications enterprise will look like are the basic questions to be answered over the next several years.

During this same time period the information industry has been undergoing its own explosive growth. Chip speed and storage technology have brought the capability to have multi-terabyte sized databases, even to small- and medium-sized companies. Our understanding regarding the value of and best design for data warehouses has evolved from earlier concepts of Executive Information Systems (EIS). Typically, EIS had limited availability in the business, often could not provide up-to-date information, and was generally limited in scope. In comparison, today's sophisticated data warehouses contain data from the entire enterprise and are available to a wide audience in the business.

United States Cellular Corporations' (USCC) evolution has paralleled these two industries' growth and development. USCC has grown approximately tenfold in less than five years with over 1.1 million customers as of this writing. The cellular business has been a duopoly up to now, but new competition from PCS (personal communications services) carriers is entering our markets in an attempt to leapfrog the cellular carriers with new technology and marketing force. The stage is set then for a new round of growth for the wireless business, but it is also a much more competitive round where timely actionable information and analysis will be increasingly critical.

USCC introduced its first-generation data warehouse just a few years ago. It served primarily as a gathering place for data from various diverse legacy systems. Our access and analysis tools were limited to reporting tools, and the amount of data was limited by our understanding of requirements and potential uses. The design of our next-generation data warehouse takes advantage of improved hardware capabilities so that terabytes of data will not be far off for us. But more importantly, the very objectives of the data warehouse have evolved. Now we are looking at a full fledged marketing and customer service tool which will incorporate external data as well as internal business data. More sophisticated tools will provide not only reporting but analysis that will help us to see correlations and ultimately business opportunities we had not previously identified. The improved technology also helps to make this information available on a scale and at a price that will allow us to extend access to the entire company. By extending this access we can bring the value of this new technology to our customers. We can find opportunities to help our existing customers more advantageously use their service. We can better understand what our customers want by more clearly understanding their usage behavior. The data warehouse provides us an opportunity to improve our overall service to our customers and thus improve our business.

This book provides new concepts in identifying and providing business value through the use of the data warehouse tool. The focus is not on technology but what business value can be brought to the enterprise with that technology. The advanced data warehouse built on these concepts and principles, combined with the advanced and affordable technology now available, provides a new and powerful tool to the telecommunications industry as it faces the challenge of competition.

James D. West
Vice President, Information Services and CIO
United States Cellular Corporation

Preface

WHEN YOU PUT a book like this together, many people who never hope to see their names in print get involved and provide a lot of help. I would like to give credit where credit is due and acknowledge those people here.

First and foremost, at least half of the credit for this book needs to go to my wife, Brigitte Kilger-Mattison. Brigitte was responsible for editing all the material, creating all the graphics, and coordinating all the efforts of everyone else involved in this project. This book could not have been completed without her painstaking attention to detail, her dedication, and her loyalty.

I would also like to thank the two people who reviewed this work as it was in progress, and who provided me with valuable feedback. Jim West, CIO, US Cellular and Bill Schmarzo, Marketing, Sequent Computer Systems. Thanks guys.

The latter part of this book, which holds the information about data mining tools, could not have been completed without the support of the software vendors themselves.

From A.S.A. (ModelMAX): Donna Bartko
From Cognos (PowerPlay): Barry Jones
From MapInfo: Tom Holec
From Seagate Software (Holos): Elvin Monteleone, Jim Jacuboski, and Judith Vanderkay
From SLP InfoWare (STATlab): Jean Schmitt and Emma Alloyer
From SPSS: Stephen Deng and Meg Card

Thank you, Michel Jambu, for your contribution to this book and for sharing with our readers some of your experiences at France Telecom - CNET.

I would also like to thank my coworkers and associates at Sequent Computer Systems for putting up with me while I pounded through this material. My thanks go to Peter Dunning, Brad Bourne, Ralph Vinci, Michael J. Notaro, Dave Klose, Debbie Koenig, Tracy Montello, Nancy Sauser, Mary Hazuka, Dave Thron, Ryan Trier, Dale Taylor, Dean Zwikel, Barry Krasner, Mabel Lattimore, Ron Shulkin, Al Polich, and the rest of the Chicago team. Thanks for your tolerance through it all.

Chapter 1

Everything's up to date in Kansas City

> *"Everything's up to date in Kansas City.*
> *They've gone about as far as you can go.*
> *They've even got a telephone invented by Alexander Graham Bell*
> *Which lets you talk to almost everyone you'd want to know."*
> —Paraphrased from a song in the musical *Oklahoma*.

THE HUMOROUS LYRICS of this song, sung by a cowboy who had just visited the metropolis of Kansas City, and the many wondrous things he had seen, provides us with an interesting perspective on just how much our lives have changed—as individuals, as a society, and as businesses—and the pivotal role that the telecommunications industry has played in making those changes happen.

It is almost inconceivable that it was only 120 years ago when Alexander Graham Bell invented the first functional telephone, a device that allowed two people to talk to each other across a stretch of wire only several yards long. In almost no time at all, we find ourselves living in a world where things that would have been considered the ramblings of madmen, or at least the wild speculations of science fiction writers and story tellers, have become the accepted realities of billions of people.

The concept of Dick Tracy's wrist radio, a device small enough to attach to his wrist but still functional enough for him to see and talk to anyone anywhere in the world, was thought to be impossible only a few years ago. Yet today, we carry cellular phones in our shirt pockets that provide 90% of that functionality with devices that weigh in at just a few ounces.

The impact of telecommunications on our lives cannot be emphasized enough. And yet, there is so much of it, and it is coming so fast, and its impact is so insidious, that we really cannot successfully appreciate exactly how far reaching it is.

People are beginning to realize that telecommunications is a lot more than simply, "… a really nice way to stay in touch." While that concept is important, and the personal connectivity that telecommunications provides has certainly proved to be a boon to the quality of our lives, this personal connectivity is proving to be only the tip of a much larger iceberg.

No one can deny that the business of telecommunications is a big business. No industry has grown so fast (remember, the telecommunications industry was born in 1847), has had such a huge impact on people's personal lives and life styles, has saved as many lives, has enabled as many deals, or has provided as much impetus for social, psychological, economic, political, and financial change as this one. It is clearly a modern miracle.

Equally clear is the fact that working within a telecommunications firm can be exciting, challenging, and more than a little bit frustrating. There is a price to be paid if all of these miracles are going to be delivered, and it is the employees of the telecommunications firms—the CEOs, CFOs, CIOs, accountants, customer service personnel, engineers, sales people, marketers, and managers—who are charged with the responsibility for turning the potential capabilities that telecommunications technology offers into a reality for everyone's life.

The purpose of this book, in the broadest sense, is to talk about that process, to develop a better understanding of it, and to try to identify those situations where the application of a different kind of technology (computer technology) most effectively supports that bigger mission. We want to talk about ways to harness the power of computer technologies (specifically data warehousing and data mining) and see how they can supplement the process of turning the technological telecommunications potentials into economic and practical reality for people, businesses, and governments all over the world.

That is, after all, the real point of any telecommunications company. We want to talk about how we can greatly enhance the process of turning ideas into actions and turning those actions into the results that we have all come to expect from telecommunications firms.

1.1 The current industry composition

While details about all of the different faces, phases, and types of telecommunications firms could take up a book in and of itself, we will spend a little time just getting a handle on what kinds of companies we are dealing with today.

The following list provides us with a reference to some of the more prominent segments of the telecommunications industry today. As is common in any high-tech industry, telecommunications is inundated with three-letter acronyms (TLAs):

- *Local exchange carriers* (LECs)—Geographically limited providers of traditional phone service to residences and businesses.
- *Specialized common carriers* (SCCs)—Companies who will contract to carry traffic between two points or provide other kinds of specialized services (a.k.a. the interconnect industry).
- *Long-distance carriers* (LDCs) or "long lines" companies—Companies that concentrate on traffic between countries or other long distances.
- *Value added carriers* (VACs)—Companies that provide compression, formatting, or other ministrations to enhance the telecommunication service.
- *Private branch exchange* (PBX)—Private phone systems, usually belonging to companies, buildings, or hotels, which manage all interoffice or inter-room communication while at the same time connecting to outside "public" services.
- *Data transport services*—An incredibly huge and almost invisible market concerned with the movement of data between businesses, including MAPI (messaging application program interface) and ECI (electronic commerce and banking).
- *Wide area telephone services* (WATS)—1-800/1-900 (U.S.) or other toll free number services that allow callers to enjoy special billing arrangements (toll free calls or calls for charges in addition to the phone charges).
- *Coin phones*—Pay phones support networks.

- *Paging services*—Devices that allow people to call a number and send a paging signal to the receiver's device.
- *Air, marine services*—Specialized services for over-water or in-the-air phone services.
- *Cellular*—The first generation of mass distributed mobile, personal phone devices.
- *Personal communication services* (PCS)—The second generation of lower bandwidth cellular type devices currently being deployed in many countries.
- *Broadband services*—A recent addition to the telecommunications arena, these cable-to-the-home services are attempting to combine cable TV with telephony to form interactive home telecommunications capabilities of extremely great depth and breadth of services.
- *Personal communication features*—Call forwarding, voice mail, remote fax, speed dial, and a slew of personal communications enhancement.

This dizzying array of industry segments, when taken as a whole, make up what we know as "the telecommunications industry." But the whole is much more than the sum of its parts in this case.

1.2 Why is telecommunications so BIG?

While the humble process of making it possible for Aunt Mabel to talk with Uncle Billy from opposite sides of the world is nice, it just does not seem like such a strong and driving force as to make the telecommunications industry as big, powerful, and important as it is. Clearly, there is something else involved.

That something else is business. While telecommunications has made personal communication available, it is for the most part a wonderful convenience, *not* a necessity. But when you look at telecommunications and the role it has played in the development of business, and the incredible role it is getting ready to play in the business of the future, you begin to get a much better idea of what the telecommunications and information revolutions are all about.

Think about it for a minute. How has business changed over the past few decades, and what has the role of telecommunications been in making that change possible?

Businesses are bigger, more efficient, more global in their perspective and more dominant in all aspects of our lives. But how did businesses get that way?

The answer is through the application of the two big technologies of the 20th century, telecommunications and computers.

The stark reality is that telecommunications, in combination with computer innovation, is making it possible for individuals and companies to function at levels never before possible. Think about the average business person's work day. How much time is spent on the telephone? How dependent is the business on the fax machine, electronic mail, videoconferencing, and conferenced phone calls? What business person can survive without a pager, a cellular phone, and voice mail?

Taken from the big picture, what does this dependence on telecommunications represent? In the simplest terms, it provides efficiency, incredible efficiency never before imagined. Telecommunications capabilities allow companies to coordinate the activities of thousands, even millions of people located anywhere throughout the world.

It makes it possible for the best minds to be applied to the most critical problems without concern for where the person happens to be physically located.

It enables people to work at levels of efficiency that are staggering by the work standards of only a decade ago, and it makes it possible for everyone to concentrate on the most critical issues that drive the business: efficiency, speed, accuracy, responsiveness, and completeness.

1.3 Telecommunications: the major driving economic force of the 21st century

In fact, telecommunications, and the telecommunications industry, has established itself as the driving, pre-eminent force to be reckoned with for business survival in the 21st century. No corporation or government agency of import is ignoring the importance that telecommunications plays.

Not only is the business' own ability to do things within its own walls being enhanced, but the capability of the business to reach the customer, to figure out what the customer needs, and to get it to him or her more quickly, is also driven by these same capabilities. The Internet interactive video, shop at home networks, home shopping services, telemarketing and direct marketing initiatives, all drive from the telecommunications space. Telecommunications is making more products available to more people than ever before.

Look at the pervasive nature of credit cards! What about automated teller machines (ATMs) and instant cash through banking cards. None of these would be possible without the availability of robust, secure telecommunications capabilities.

The business' ability to work and function efficiently with other businesses is also being enhanced dramatically. Electronic data interchange (EDI) and other related technologies are making it possible for companies to work together and to coordinate their activities through the simple addition of a telephone line to the equation. The "just in time" manufacturing paradigm of the 1980s and the efficient customer response (ECR) model of the 1990s are both telecommunications-driven capabilities.

Banking, medicine, government, and almost every other industry are being affected by the telecommunications industry in similar ways. No one anywhere is untouched.

1.4 Knowledge management enablement—the biggest factor of all

What all of this activity in telecommunications has really done is create an entirely new world, a completely different world where many of the limiting factors for individuals, businesses, and governments are being swept away. We are no longer limited by space, time, and distance the way we used to be. We are no longer crippled by miscommunication, ignorance, and inefficiency in the ways we used to be. In fact, we have "raised the bar" for business and for life in general.

The new operational paradigm, the one created in no small part by the telecommunications revolution, has been to shift our concerns from the management of things to the management of knowledge itself. Raymond W. Smith, CEO of Bell Atlantic, in a paper he published in the *Annual Review of Communications* (IEC - 1994-1995) called "Loaves and Fishes Revisited: What's Driving the Telecommunications Revolution?" writes:

> Technologies have enabled "knowledge" to take the place of raw material as the most valuable ingredient in any given product. Unlike physical inputs, knowledge is a resource you cannot use up. The more you dispense in your organization, the more you generate. Just like the biblical loaves and fishes—no diminishing returns, only expanding ones.

When you look at what is really driving business today, and when you look for reasons why telecommunications is as pervasive as it is, this explanation makes the most sense.

1.5 The ultimate environment

Ultimately, what we can envision is a world where all companies and all people are connected through a vast network of differing capabilities: personal communications, mail, electronically managed purchases, electronic delivery of entertainment, information, and time-saving features that go far beyond what is possible today. But we have a long road before we get there.

1.5.1 Failed excursions into the new frontiers

Not all of the history of telecommunications is paved in gold and glory. There have been many bumps along the road. Besides the obvious and ever present demands to mix a competitive environment that drives change and a monopolistic environment that promotes low cost and consistency, we also have the problems of defining what the new products and services will be and how they will work.

Many "experiments" have failed. Some are still being tried. Some have resulted in big success. Almost all of them involve the merging, partnering, or cooperation of many large firms.

Pacific Bell is currently undergoing the investment of over $15 billion in the wiring of households for broadband capabilities. Will this prove to be a prudent investment?

AT&T has announced a major infrastructural overhaul in attempts to integrate recently purchased McCaw Cellular with its already exhaustive environment.

The recent announcement of British Telecom and its assumption of full ownership of MCI creates some interesting competitive possibilities on an incredibly grand scale.

Bell South has recently experienced less than overwhelming results in its attempt to expand fiber networks into the homes of Hawthorne, Florida.

GTE's integrated network experiments in Cerritos, California have left much to be desired.

US West/AT&T/TCI were unable to provide an integrated network environment in the Denver, Colorado area.

Internationally, titanic telecos are merging in order to penetrate new markets and to provide more comprehensive services to customers.

Hong Kong Telecom (as a part of the Cable and Wireless Federation) is participating with firms like Mercury (United Kingdom) and Tele2 (Sweden) to expand into other markets. These markets include Jamaica (where 80% of the infrastructure is owned by Cable and Wireless), Latvia (where 75% of Tilts, the former government telecommunications service, is now operating), and China (where AsiaSat1 is owned jointly by the Chinese International Trust and Hong Kong Telecom).

Clearly, if you are in the telecommunications industry today, then mergers, cooperatives, and other types of deals are a big part of a successful long-term strategy.

1.6 Future directions

If everything in the telecommunications industry were to stay frozen in place as it is today, then there would still be a need for the application of the kinds of technology that we will be talking about; but the fact of that matter is that the industry is still growing, and growing at a phenomenal rate.

Some of the bigger initiatives include the following:

- *Information superhighway*—The movement to create a worldwide support network that will have the capability to connect everyone in the world on a massive scale for the movement of conversation, data, information, and commerce.
- *Broadband*—The incursion of the traditionally entertainment-based media into the telephony and data transmission space.
- *One phone number*—The drive to give each person a personal phone number, and to have the number respond to whatever physical locations or devices that person has available. A version of this service is already being offered in the United States, where an unanswered phone call to your home will switch to your cellular phone, and then perhaps to your pager or voice-mail system.
- *Electronic commerce*—Electronic banking, ordering, delivery, scheduling, and the movement of data between businesses and customers.
- *The Internet*—Providing everyone access to the World Wide Web (WWW).

1.7 Telecommunications and technological innovation

One of the most fascinating things about the telecommunications industry is the way the business has been able to take advantage of technology, data, and knowledge in order to provide better service to its customers.

We can see the "symptoms" of things to come in some of the very earliest days of telecommunications. From the very outset, telecommunications firms had a desperate need for hard data and the ability to interpret it.

1.7.1 Peg counts

One of the first "data warehousing/data mining" type applications to be used in the industry is evidenced in the practice of maintaining "peg counts." When the first telephone exchanges were created, all connections between people were made by hand. Operators plugged one phone into another via a "switchboard." In order to gain some idea of the flow or volume of calls that the system was carrying, early phone operators kept track of all traffic on a peg board. This was a simple wooden board with rows of holes in it. Each time a phone connection was made, the operator would move the peg one notch forward, thereby increasing the "peg count" for that exchange.

The term "peg count" is still used today to indicate what the volume of traffic through an exchange might be, but very few of the people using the term really understand its origins.

1.7.2 Business drives technological innovation

Another interesting bit of telecommunications folklore can be found around the invention of the first automatic phone-switching device. In the earlier days of telecommunications, operators connected all calls. The folk tales state that an undertaker by the name of Almon B. Strowger, who was, coincidentally enough, from Kansas City, became convinced that the local exchange operators were purposely routing all calls for undertaking services to his competitor down the street. He swore to develop a mechanism that would make these operators obsolete and eventually invented the Strowger switch, a device that allowed for the first automated connecting of two telephones by mechanical means. The first Strowger switch was put into use in LaPorte, Indiana, and it was eventually adopted by AT&T as the de facto standard.

1.8 The three strategic options

Ironically then, we have in telecommunications an industry that almost single-handedly spurred the creation of a revolution in the conduct of business itself, and yet, as is so often the case, we find that telecommunications firms can often be the last to take advantage of these capabilities for themselves.

What is clear, however, is that these firms are learning quickly and are in many ways surpassing their brothers and sisters in other industries in their ability to figure out how to make the knowledge they possess more available to more people.

Of course, if we want to talk about a better way for telecommunications firms to meet their strategic objectives, then we must start with an understanding of what those objectives are.

At the highest level, there are only three strategies open to any business that wants to achieve dominance in their market place. They must either (1) excel at being the best low-cost provider of services, making them the volume market dominator; (2) be superior at the quality of what they are doing, distinguishing themselves as the best provider of services; or (3) become the best marketers of service, pursuing what is known as "customer intimacy," and becoming the company that responds to customer needs and wants better than anyone else (known as attempting to dominate the customer "share of wallet").

Obviously, the strategy that a particular firm is pursuing will dictate how data warehousing and data mining can help them achieve these ends, but in all cases these technologies represent one of the best ways to accomplish any of them.

1.9 Customer intimacy—from "network is king" to "customer is king"

One of the general trends in telecommunications has been a shift in the emphasis that most companies place on the role of engineering and infrastructure versus the role of marketing. In the government-sponsored monopoly days, efficiency, size of infrastructure, and efficiency of operations were the keys to success. With deregulation, however, comes a change in emphasis from a network-is-king philosophy to one that emphasizes the role of the customer.

1.9.1 Marketing as the driving force

In companies that decide to take this approach, marketing and a better understanding of customers becomes the number one imperative. And for these companies, data warehousing and data mining provide them with the ideal means to exploit it effectively.

One tremendous advantage that any telecommunications firm has over its competition and over other industries is the fact that the telecommunications firm knows more about its customers than anyone else. They know who they are, where they live, where they go, and what they do. Keeping track of customer activities is a byproduct of the service they provide, and figuring out how to capitalize on that knowledge is what data warehousing and data mining is all about.

Marketing databases, customer information systems, enhanced customer service capabilities, predictive behavior models, and integrated marketing strategies are just a few of the tools that the telecommunications firm can use to gain a dominating control over its relationship with customers. And this kind of customer loyalty is not easily stolen by competitors.

The key data stores to support these kinds of activities are customer and transaction based. And the key activities that make exploiting it possible are all marketing based. We will spend a significant amount of time throughout the rest of this book talking about these activities, and we will show you how they can be exploited quickly and economically.

1.10 Operational efficiency—being the low-cost provider of choice

If a company decides that it wishes to be the best low-cost provider, then its biggest challenges are going to arise in trying to figure out how to best coordinate, measure, and control the many different functions that drive the delivery of services to the customer.

In these cases, data warehouses that emphasize monitoring of the company's "value chain" and coordinating the activities of different groups of knowledge workers will be the keys to success. Many of the infrastructural and organizational issues that we will be considering focus on how to put these kinds of operational monitoring and control systems together to the benefit of the entire firm.

1.11 Technical proficiency—being the best at what you do

For the firm interested in becoming the technological leader, the biggest source of problems will probably come from the challenges that constantly changing innovation and positioning create. A company that wants to stay on the leading edge of technology needs to be aware of what all of the pieces of the organization are and what they are doing. Value-chain-driven warehouses meet these needs.

At the same time, technical proficiency requires the optimization of several distinct operational areas. In these cases, analytical warehouses and advanced mining tools can provide engineers and managers with the information necessary to drive their business ahead.

1.12 Conclusion

In summary, it is clear that telecommunications companies have been responsible for the creation of a new and exciting business environment, as well as a significantly large and successful industry in its own right.

This chapter has provided us with a better understanding of the scope, nature, and major issues that drive the telecommunications firm today, as well as with some ideas about where they will be headed in the future.

In the following chapters we will be considering the telecommunications company itself in far greater detail. We will examine ways that the company, the information systems, and the infrastructures can be better utilized to improve profitability and competitiveness to even greater levels of accomplishment.

Chapter 2

Why warehousing and how to get started

So, given the much better understanding we now have of what the telecommunications industry is about, and the rich history of innovation in computer systems and decision support development that this industry has enjoyed, we are ready to settle down to the complicated business of trying to figure out how to apply these latest technological advances, *data warehousing* and *data mining,* to the business of helping telecommunications firms to survive and thrive in an ever more competitive marketplace. In order to do that, we need to begin by answering some fundamental questions.

These questions include the following:

- What is data warehousing?
- What is data mining?
- What is our objective in using them?
- Why do we believe that these approaches can provide significant value?

If our answers to these questions make sense to you, then you will know that your time spent reading this book is going to be of value. If the answers do not make sense, then data warehousing and data mining may not be the kinds of technological solutions you should be considering.

2.1 Background of data warehousing

Dozens of books and hundreds of magazine articles and white papers have been written about the subject of data warehousing and data mining over the past several years. Data warehousing has become the *hot, new, latest and greatest* technology to come along since the invention of the database.

But what is a data warehouse, really? If you have spent any time at all talking with hardware or software vendors or reading magazine articles, you will have discovered pretty quickly that, while everyone thinks that these are great things to do, no one can agree on exactly what they mean or how they should be done.

This, of course, presents us with an immediate problem. How can we possibly write a book that claims to tell you how to apply data warehousing and data mining to the telecommunications industry when we don't even have good definitions for them.

2.1.1 The history of the data warehousing phenomenon

Data warehousing is perhaps one of the strangest phenomena to come upon the corporate data processing scene. It is an approach that involves no specific discipline, no science, no clear rules or guidelines, and no solid set of tangibles by which you can describe it. It is more like a philosophy or a way of looking at things than it is a true data processing enterprise.

Yet, there have been dozens of books and hundreds of magazine articles written about it, and millions of dollars spent on promoting it, by hardware vendors, software vendors, conferences, trade shows, and all sorts of dignitaries and gurus. It is a phenomenon that has become all too common in the data processing industry.

Ultimately, the roots of data warehousing can be found in the disciplines of database and data management. For years, in fact since the early 1960s, organizations and theorists have realized that managing information, and the data that makes that information usable, was the driving force behind modern large-scale corporate business enterprises. Unfortunately, data and information are not easy things to manage.

In the 1950s, the first major attempt to provide for the wholesale management of data was attempted. Specialized software, called database management systems, was created to help businesses get control over the vast landscape of data that they were being forced to manage. These early databases (like IBM's IMS or Cullinet's IDMS) were complex, large-scale data management environments that made it possible for developers of business systems to capture, manage, and efficiently make available massive amounts of data. (Some databases were as large as several hundred megabytes in size.)

Of course, as time went on and as technology got more complex, these "old fashioned" databases proved to provide less flexibility and responsiveness than businesses needed. Then, in the mid-1960s came the invention of the concept of a relational database. Relational databases were different than these earlier versions because they allowed people to gain more access in more ways with much less dependence on programmers. These relational databases took the industry by storm, to the point where almost no other types of databases even exist anymore.

While everyone was still reeling from the shock waves that the relational database revolution created for the developers and users of business systems, we found ourselves in the throws of yet another revolution. This one involved new kinds of hardware (large-scale, "open" UNIX systems and personal computers) and a new way of looking at system development. This round was called the "client-server" revolution. Suddenly, people found themselves with powerful personal computers on their desktops. These computers had more power in them than the older mainframe systems used to have back in the early days of data processing. Not only were they powerful, but they allowed the business person to become his or her own programmer and database administrator (DBA). Users built large, sophisticated systems using tools like LOTUS, Excel, Access, or dBASE-IV. The result was a major shift in our comprehension of how information could be managed differently.

One good thing that came out of the client-server revolution was an understanding of what the true potential of the combination of dynamic networks, low-cost UNIX database servers, and intelligently loaded personal computers might be in order to change the shape of the corporate working environment. The bad thing was that we discovered, once again, that the management of the data that drove these processes became the weakest link and the most limiting factor, keeping us from realizing the full potential that was being offered.

It is upon this scene that the data warehousing revolution presented itself. You see, in order to truly capitalize on the potential power that these technological revolutions were offering, we needed a new, improved, more robust and more business-oriented way of looking at data processing than we ever had before. The old way of viewing systems as online transaction processing systems, or reporting systems, or consolidation systems, all failed to let us organize things in a way that really put together all of the pieces that the last several generations of innovation had been delivering to us.

This is why, when you look at it on the surface, data warehousing seems to be such an amorphous, undefined, intangible concept. It seems that way because it has to be. It seems that way because it is really an approach that tries to take all of these capabilities, learn from all of the failures and problems we have experienced in the past, and put them together in a way that makes sense out of it all.

2.1.2 Data warehousing—in a nutshell

The fundamental principle that warehousing presents is simple, but in many ways is revolutionary.

Every theory of data management before data warehousing was driven by several fundamental principles. For over 40 years the industry has been a slave to these principles. What is truly revolutionary about warehousing is the way that it discards or drastically alters the attention we give to these principles.

2.1.2.1 The "science" of data management the "old fashioned" way

The core concepts that have for years dictated the rules of delivery for DBAs and system developers have been these:

1. *The elimination of data redundancy and the minimization of disk storage space*—Everybody who was anybody agreed that the primary goal of any data management exercise was to try to figure out how to minimize the amount of data that was being stored. It was never, or hardly ever, permissible to store the same data element (name, address, sales-code, etc.) more than once. A successful activity was one where hours, days, and sometimes even months were spent trying to figure out all of the ways that people might want to use data, or all of the theoretically proper ways to store it, and then go ahead and store it that way once and for all.

2. *The use of entity relationship and normalization modeling techniques—*These two techniques became the hallmark of the database design process, and both techniques demand that a theoretical discipline (not a business-oriented one) determine what goes into the database. Naturally, the logic that drives such systems is one designed to

maximize the construction of large corporate online transaction processing systems.

3. *The dependence on the systems development life cycle and JAD (joint application development) sessions as the means of designing systems—* These techniques represent the sum total of several decades of experience in the building of computer systems. Unfortunately, like the database design disciplines mentioned above, they too assume that you are building a type of system that is not part of the new organizational paradigm that data warehousing systems require.

2.1.2.2 The principles behind the data warehousing paradigm

As we said, the principles behind data warehousing only begin to make sense against the backdrop of what the old principles of system design used to be.

Where the old paradigm made the elimination of data redundancy and the optimization of disk storage space the key to design, data warehousing says that the duplication of data is okay, and in many cases a good thing to do.

Where the old science of data management insisted upon the use of entity relationship, normalization, and other "theory-based" approaches, data warehousing leans to a more practical, results-based approach. The Star Schema design paradigm replaces the older approaches.

Where JAD sessions and the systems development life cycle provided a bureaucratically bound, structured way of looking at the relationship between users and computer systems personnel that made collaborative and creative systems development almost impossible, data warehousing assumes an iterative, interactive, learn-as-you-go approach to systems design.

2.1.3 What is a data warehouse?

For want of a better description, in general, a data warehouse is a collection of data copied from other systems and assembled into one place. Once assembled, it is made available to end users, who can use it to support a plethora of different kinds of business decision support and information collection activities. That's it. It's that simple.

A data warehouse is a collection of data assembled for the express purpose of making information available to users.

Of course, while the concept is simple and the imperative objectives are clear, the execution of an effective warehouse is obviously a lot more complicated. We will spend a significant amount of time discussing those kinds of issues.

2.1.3.1 What's important about warehousing?

What's important about warehousing, however, are *not* those details. What is important is the fundamental shift it represents in the way people view the systems development process because of it. Warehouses are collections of information assembled for users to meet their practical business needs. It is not about theory, and it is not about computer systems. It is about business needs and the survival of the corporation in a competitive environment.

Of course, once the data is assembled, we need to come up with a way to make it accessible to end users. It is in this area where yet another revolution makes the whole warehousing paradigm even more powerful, and that is in the area of *data mining*.

2.2 Data mining

While the data warehousing revolution addresses the needs of business people by making data conveniently accessible, the complementary revolution, the data mining revolution, makes it possible for them to do much more with that data.

For years, the science of accessing and making intelligent use of information has been the darling project of hundreds of theorists in the areas of psychology, computer-human interaction, statistics, artificial intelligence, decision support, and executive information systems. They have been chipping away at the problem of how to empower business people and help them to make smarter, more informed decisions. Finally, the advent of new, extremely high-powered personal computers and scientific workstations has made it possible to deliver these capabilities to the business person's desktop at a reasonable price.

The field of data mining, like the field of data warehousing, is rife with conflicting definitions and controversy. Our definition of mining, like warehousing, will remain generalized, simple, and functional.

We include within the category of data mining any and all tools that, when provided to end users, give them the ability to do ad hoc and/or user-defined analyses of information in order to solve specific business problems.

Included in our repertoire of tools are the following major categories:

1. *Agents*—These are software constructs that allow users to specify the kind of information they want to access and then empower the computer system itself to go and find it, analyze it, and report back about

what it found. Agents are usually built into other types of products, but some very sophisticated agents have started to show up on the desktop and the Web.

2. *Query and reporting tools*—These products represent the latest generation of ad hoc access and reporting tools. These tools have been evolving greatly since the early days of Focus, QMF, and Oracle*Forms. Today's query and reporting tools make it possible for people to ask for business information in business terms with a minimum of structured query language (SQL) and other programming language requirements.

3. *Statistical analysis tools*—These products represent the latest generation of the traditional mainline statistical report tools like SPSS and SAS. In this day and age, these products are finding new homes on the desktops of many executives that never before considered themselves to be "statisticians."

4. *Data discovery*—Into this category we place all products that apply any kind of statistical, superstatistical, or artificial intelligence to the process of interpreting large amounts of data. Techniques like CART (classification and regression trees) and CHAID (chi-squared automatic interaction detector), constructs like neural networks or decision trees, and a wide assortment of specialized tools make prediction, estimation, and forecasting a scientific as opposed to an artistic process.

5. *OLAP*—Online analytical processing, also known as multidimensional spreadsheets or multidimensional databases, provides business analysts with the ability to "surf" through their stores of data in order to hunt for new insights or conditions.

6. *Visualization*—Products that make data easy to understand and interpret by displaying it on maps, multidimensional graphs, or other innovative techniques.

7. *Web-based discovery*—A whole new class of data mining possibilities are being made possible by the new intranet and Internet revolutions.

No matter what kind of business problem you are trying to solve, data mining tools will make the job easier and more effective.

2.2.1 Why should one seriously consider using these approaches?

From an overall theoretical perspective, the preceding descriptions of data warehousing and data mining must certainly sound attractive. But if we are going to get serious about using this technology to solve business problems, we need to have a clear understanding of exactly how the technology is going to be applied.

Unfortunately, the history of the application of data warehousing and data mining has been anything but rosy. Too often, people have attempted to apply them without taking the time to get a clear understanding of what they are, how to use them, and what to do with them. The result has been hundreds of failed projects.

What has become apparent is that the successful projects are those where the people developing the system have a clear conception of business focus. If the architecture and design of the system is driven by business need and business focus, then the system can succeed. If the focus is driven by theory and data processing need, then it will fail. It is our intent to show you exactly what we mean by that and how anyone can use the same kind of approach to guarantee their own success.

2.3 Why are these approaches so exceptionally valuable to telecommunications firms?

Based upon the previous discussion, it should be clear that these approaches can be useful for any business, but why do we feel that the value to telecommunications firms is especially great? There are several reasons.

2.3.1 Data intensity

Telecommunications is one of the most data-intensive industries that anyone can imagine. The main product of the telecommunications firm is the *call* or *connection*, and customers can literally create hundreds of thousands of these kinds of transactions in any one day. The company must execute the transaction and then keep track of it in order to monitor network performance, issue bills, and perform network planning and optimization exercises. In an environment with so much raw data to deal with, the data warehousing and data mining paradigms make a lot of sense.

2.3.2 Analysis dependency

Telecommunications firms are incredibly dependent on their ability to wade through tons of raw data and make intelligent decisions based upon what they find. As opposed to the traditional manufacturing firm, which has a lot of physical inventory to track and well-defined procedures to measure how well things are going, the telecommunications firm has no tangible goods to track. Its product is a connection or a session. Therefore, these firms are severely dependent on raw, abstract data in order to generate bills, measure network effectiveness, and in other ways run their business. Because of this, telecommunications firms are highly dependent on raw data analysis. And this is what data mining and data warehousing do best.

2.3.3 Competitive climate

The history of telecommunications firms takes it back to the "good old days" of government regulation and enforced monopoly. Firms that are created in this environment develop with a serious lack of foresight in terms of understanding how to function in a competitive marketplace. Because of this, telecommunications firms are only now beginning to learn how to create the technological infrastructures (marketing computer systems) and organizational cultural preferences (market and customer-driven thinking and activity) necessary to be effective in the new competitive environment they find themselves in.

Data warehousing and data mining are ideal tools to help address these inherent weaknesses. A wide range of tools and approaches from this area help companies develop new marketing-based infrastructures quickly, and the new tools help empower organizations to make the cultural changes necessary to make them more customer-centric and less technology-centric.

2.3.4 Technological change at a very high rate

The telecommunications industry is perhaps the most chaotic and difficult to work in from an engineering and infrastructure perspective. Almost quarterly new innovations in switching, networking, and transport media force telecommunications firms to constantly re-evaluate their investments in infrastructure and design of networks. In an environment like this, it is critical that the organization develop the means to adjust to changes quickly and to analyze how well the changes are being implemented.

A data warehouse is a well-suited approach that can respond flexibly and quickly to this kind of tumultuous organization.

2.3.5 Historical precedent

And finally, we find that the telecommunications industry in general has a long and rich history of making use of data management innovations of this kind to their advantage, even from the earliest days. The rich tradition of complex data analysis that has made telecommunications firms as efficient and profitable as they are is an ideal starting point for the development of a powerful mining and warehousing organization.

2.4 Organizing the process

So, we have an industry whose major characteristics are that it is big, incredibly profitable, increasingly competitive, almost unbelievably complex, technically sophisticated, and provides critical services to the consumers, businesses, government, and industry around the world—and a technological approach that seems an ideal solution to a lot of the problems they face. How then do we go about figuring out how best to harness this technology in an efficient and profitable manner? How do you figure out how to get started? As large and complex as the computer systems are that drive telecommunications companies, and as contradictory and complicated as the business organizations are, and with an environment as chaotic as their's, it can be a daunting task to come up with a plan for making these benefits available. In order to deploy this technology, we have to know several things, as described in the sections that follow.

2.4.1 An inventory of the existing computer systems and other technological infrastructure

If the data warehousing solution that we are proposing is going to make sense for your business, then the first thing you need to do is to figure out what your existing information systems are doing and what kinds of information they are already managing. An effective solution is one that leverages as much of the existing "legacy systems" environment as possible. The warehouse needs to hook up to the legacy systems as cleanly as possible, and with the cost of computer information systems as high as they are, the more of it we can leverage the better. Therefore, the first step must be an inventory of existing systems.

It is here that many of the data warehousing approaches really fall short. Most approaches take a theoretical "tabula rasa" approach. These theorists start with the basic premise of "assume that you have no existing systems to worry about." Our approach will start with exactly the opposite tack.

2.4.2 A roadmap and an approach for how to deploy data warehouses in general

After developing a good sense, from a technological perspective, of what the existing information systems contain and how they are being run, we can then turn to the business of figuring out how best to assemble a data warehousing environment that makes the best use of those legacy systems. We will need to put together a template for the construction of such a system that takes advantage of the company's existing infrastructure and makes it possible to incorporate new technologies to the best advantage. Included in this roadmap will need to be an approach for figuring out what to put into the warehouse and when.

2.4.3 A roadmap for understanding how to diagnose and develop a plan for identifying the best things to put into the warehouse and which data mining tools to use

Of course, after we figure out how we are going to build the warehouse, we then need to address the issues that come up when trying to determine the kinds of things that the warehouse should do. For this we will have to turn to one of the newest "sciences" of management and data processing, referred to as the *knowledge management discipline*. Through the knowledge management approach we will define a process of figuring out what kinds of things the warehouse should hold, defined it in business terms, and then propose a method for defining the kinds of data mining tools that will best be used to those ends.

Finally, we will be able to put all of these inputs together to develop a plan for the construction of our complete warehouse.

In the following several chapters we will consider in some detail the issues, roadmaps, and approaches we have been discussing. By the time we are through, you will have been provided with all of the information necessary to help you develop your own plan for developing a telecommunications warehouse.

Chapter 3

The knowledge management view of business and warehousing

So far we have spent a lot of time talking about the telecommunications industry itself, its history, its foundations, and its reasons for existence. We have spent very little time talking about computers, data warehouses, and other kinds of technological issues. There is a very good reason for us to have started out this way and to continue on this track for just a little while longer.

Our reasons should by now begin to be clear. The history and much experience-based analysis of data warehousing have shown us unequivocally that the deployment of successful data warehouse and data mining initiatives, in any industry, occurs only when the system that is being implemented is derived from and delivered to the business in a way that is industry specific and tailored specifically to the needs of the business for which it is being created.

There are many reasons why this is true, and the biggest reason should also be the most obvious. Effective warehousing solutions are deployed to solve specific business problems. Therefore, only knowledge of how the business works and what the business people need to do their jobs more effectively can provide us with the direction necessary to make good warehousing decisions. In other words, if you have the needs of the business in mind before you start designing your warehouse, then the chances are much higher that you will be successful in your efforts.

Now, that statement may seem to be rather trivial and nonessential when you think about it. Of course the warehouse should only be built with the needs of the business in mind! And of course a successful project will be one that has the business' needs in mind before you start building it. But how many books or lectures about data warehousing and data mining have you seen where this is the first and most important part of the conversation? Hardly any.

In reality, what happens in the vast majority of cases is that "experts" in the area of warehousing will either tell you that the particular industry is not important and that the same warehousing tools, techniques, and approaches will have equal value for everyone, or they allow you to assume that they know your industry and your business and that they have it in mind before they start (a dangerous and false assumption in far too many cases).

3.1 The knowledge management revolution

When you think about it, our insistence on developing an understanding of the business of telecommunications before discussing the technology of data warehousing is a fairly radical starting point.

While we don't think that it is necessary for someone to have 20 years of experience in telecommunications before they can provide any value to the discussion of warehousing in the business, we do believe that there are many things unique to the industry that anyone attempting to approach warehousing in this field should be aware of and should be ready to address.

There is, in fact, a brand new and relatively radical management science/information systems discipline, which is starting to get a lot of people's attention, that really addresses these kinds of issues head on. It is called knowledge management, and we are going to spend the rest of this chapter getting to know exactly what knowledge management is about, how it applies to data warehousing and data mining in general, and how it can be harnessed to help us define how warehousing and mining can be done most effectively in the telecommunications space.

3.1.1 Knowledge management principles

The basic framework for the understanding that knowledge management creates is really a pretty big one. It addresses not only computer systems development theory, but actually tries to help business managers come up with newer, more effective, and more realistic ways of defining and understanding the process of running a business efficiently. The underlying thinking goes something like this.

The whole theory of business in the 20th century has been based upon the same core model of what a business is about. It basically views businesses as collective groups of people and processes, brought together in order to convert raw materials into usable goods. The "assembly line," the "economies of scale," and the "division of labor" are the key driving paradigms that model the entire business process for us.

In knowledge management terms, we would say that specialization or fragmentation of processes is the way that businesses attempt to accomplish efficiency. So, what we find is that most very large corporate organizations are made up of large groups of specialists. Marketing, engineering, sales, and finance people, all concentrating on optimizing their piece of the huge puzzle that makes up a corporation's organizational structure.

In the non-knowledge management world then, process and specialization are the keys to understanding business.

The knowledge management advocate begins by questioning the very basic set of assumptions. While the business of the early 20th century may very well have been process and specialization based, the business of the late 20th and early 21st century is looking at it from a radically different perspective.

One of the early pioneers of the concept of knowledge management was management guru Peter Drucker. Decades ago Drucker created the concept of the "knowledge worker." He stated that more and more of the business was driven by the activities of nonprocess-oriented knowledge analysts and knowledge workers. Even at that time, he had noted that a large percentage of the nation's gross national product (GNP) could be attributed to "knowledge work" as opposed to "real work" (using the term loosely). Today, that observation is truer than ever.

According to the knowledge management advocates, the core competency of a business organization is no longer simply the processes, capital, and raw materials they can muster, but is driven first and foremost by the way they manage knowledge. A corporation is seen as a collective group of thinking and knowledgeable people using their collaborative knowledge to create benefits for customers.

This model certainly helps us explain why industries such as banking, finance, insurance, and investment brokerage wield as much economic power as they do. These are industries whose primary product is the application of knowledge to the solution of people's problems. (Insurance agencies address people's risk needs, financial institutions address their cash flow and credit needs, and so on.)

Telecommunications firms are a special form of this kind of knowledge-based business, but they clearly fall into the same category. They address people's and business' needs to communicate with each other; they have no product as such; and they have no manufacturing process to speak of.

In fact, the thinking goes, understanding your business as a knowledge transformation and focusing agency is the way for your company to gain competitive advantage in a clearly intensively competitive marketplace. The companies that figure out how to better leverage the knowledge they have, and use it to solve people's core needs, are the companies that will thrive in the 21st century.

A new pattern in the development of corporate computer systems budgets shows us a clear reflection of these basic principles. What we are finding is that corporations, in almost a wholesale fashion across all industries, are slashing their budgets for the support of traditional "kingpin" status operational support systems. It no longer makes good business sense to spend large amounts of money on the enhancement and improvement of these types of systems. This is a major shift in thinking from 10 to 20 years ago, when having the best order processing, accounts receivable, or sales tracking system was seen as the key competitive differentiator for almost all businesses.

Today, the thinking about operational systems is that you need them to keep up with or have operational parity with your competitors, but that strategic advantage comes from being better at managing the information that these systems create.

This also explains the radical and almost wholesale abandonment of extremely expensive and highly customized human resources, accounting, and manufacturing control systems that corporations used to favor in exchange for packaged solutions like Oracle Financials, Lawson, S.A.P., and PeopleSoft.

What is becoming obvious to increasing numbers of CEOs and CIOs is that decision support, data mining, and other kinds of data warehousing type applications are the real keys to their long-term competitive success.

So, if we are going to start looking at the business organization as a knowledge transformation company, as opposed to a product production company, then the next thing we need is a way to identify the kinds of knowledge that are driving that process.

3.1.2 The organizational footprint and what it tells us about knowledge transformation processes

From the business computer system's perspective, what we will inevitably find is that each specialized area of the business is going to tend to have its own collection of core operational systems (see Figure 3.1). The specialization of operational systems will reflect the specialization dictated by the business' "footprint." Accountants work with financial systems, marketing people work with the marketing and sales systems, and so on. In some industries, this footprint can become so pervasive that specialized hardware is actually invented to better service certain kinds of needs.

In almost any industry, therefore, you will probably be able to find an almost direct correlation between the basic organizational structure of the business and the core operational systems that drive them (except, of course, in those cases where radical and recent reorganizations of the business have forced a dealignment). This alignment of business functions and computer systems provides us with a clear impression of where the different clusters or collections of knowledge within the business are and how they relate to each other (see Figure 3.2).

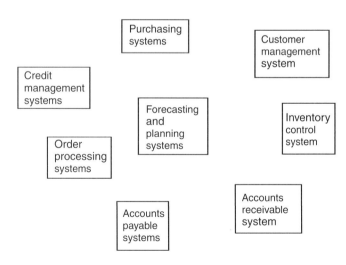

Figure 3.1 Core systems in the business.

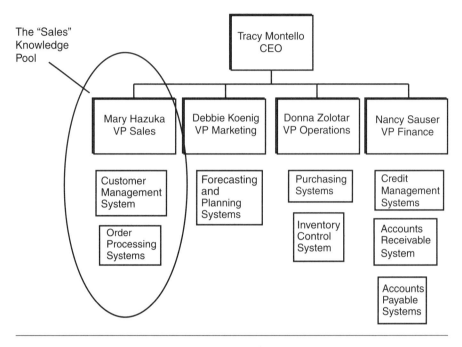

Figure 3.2 Alignment of the organizational structure and the core systems into knowledge pools.

This incredibly pervasive fragmentation of the business' overall structure sets up the very real possibility of creating organizations where the employees lose touch with what the business is really about and how their functions "fit together." In information systems development parlance, we have come to refer to the database and computer systems reflection of this characteristic as isolated "silos" of information (see Figure 3.3).

The risk (and the observation) is that the fragmentation of processes creates knowledge "sinks" that will tend to hold on to, protect, insulate, or sometimes even hoard information that might be useful to other areas of the business, but which cannot be accessed for a wide variety of reasons:

1. It might be that the information in one knowledge silo is mechanically or technologically inaccessible for any number of reasons.

2. It is possible that the people who created a certain information silo (engineering information, for example) might have constructed the knowledge storage in such a way that only they understand it (for, example, it is branded with their identity and translated into their

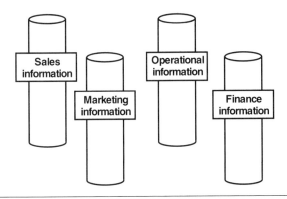

Figure 3.3 Silos of information.

professional lingo, a language which is not decipherable by someone not part of that group).
3. In some cases, business groups may hoard information because they do not want other business areas to be aware of the details about their aspects of the business.

It is actually quite ironic to note that one of the side effects of the segmentation of business functions is to in fact create large groups of specialists who will in many ways see each other as competitors or at least functional antagonists.

At one large cellular phone company we worked with, the credit department had developed a reputation for being so negative and unsupportive of marketing and operational efforts that it became known as the "antisales" department. At a small, locally based carrier's offices, the customer service department had so much trouble finding out about the status of customers' billings that they ended up creating their own duplicate copy of call detail records on their own departmental server machine so that they would be able to answer questions while on the phone with a customer.

So, as you can see, when we approach the problem of information systems and data management from a knowledge and its management perspective rather than from a data and its technology perspective, we begin to identify some inherent weaknesses, not in the information systems (which are really nothing more than a reflection of the businesses structure) nor in the structure of the business, but in our basic assumptions about how to make a company efficient and effective.

3.2 Efficiency optimization—optimize the silo or optimize the whole

It is here where we can stop for a second and recalibrate our discussion back to the issue at hand, which is namely: What has all of this got to do with data warehousing and data mining, and how can we use the knowledge to drive our strategy for implementing them?

The theory of data warehousing states that if we can make copies of operational data and place it into data storage areas where the business people can get at it easily, then we will create a new capability. We will create an environment that makes it easy for those business people to gain more and more in-depth knowledge about the business, its processes, its customers, and its operational efficiency, without having to write specific operational systems to pull all of that information together.

The good news is that this scenario is very valid, and many companies have proven it through their own use of warehousing and mining. The bad news is that the approach is too general to use in any specific way without more guidance.

Upon closer examination we will see that, from the knowledge management perspective, there are actually several different kinds of warehousing scenarios that we can envision. Some are time related and others are knowledge sink related.

From the time perspective, we can see that warehouses can help organizations view information from a historical perspective (capturing detailed historical information about what has happened in the past and making it available for analysts) or from a real-time perspective (making it possible for users to get snapshots of information about what is happening today in the environment). We call these systems passive (historical) and active (real-time) warehouses.

For example, we can provide the business with a historical database of call detail transactions and other customer "sales" information, allowing sales analysts to study the different types of customer calling patterns. This will enable sales to do a better job of serving customer needs by understanding them better. This is the historical type.

On the other hand, we can use that same call detail information to allow credit and security analysts to be made alert to possible occurrences of ongoing fraud. (For instance, these systems can check to be sure that the same customer is not making calls from Los Angeles and New York within three hours of each other. Since a person cannot be in both places at the same time, this is a strong indication that a phone number has been cloned.) This is the "real-time" type. The two types are compared in Figure 3.4.

From the knowledge sink perspective, we can see that a warehouse can also be used to help optimize a particular silo of knowledge management (we can create a sales warehouse to help people analyze and get smarter about the sales process), or we can use the warehouse to do the following:

- Help bridge the gaps between silos of information;
- Help business people to defragmentize the business operational perspective;
- Help the different specialized areas to make more coordinated decisions that add value to the overall process of doing business, as opposed to optimizing individual functions.

Alternatively, we might want to create a system that keeps track of all contacts that a customer has had with our company:

- With billing (do they pay their bills and how much do they pay?);
- With customer service (how many and what kinds of complaints have they registered?);
- With marketing (what kinds of programs have they been offered?);

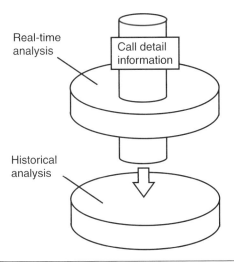

Figure 3.4 Call detail, a single silo warehouse (real-time vs. historical).

- With call detail (how much are they using our services?).

This kind of integrated view of the patron can provide the customer service department with valuable information that can drive the whole customer satisfaction and rating process (see Figure 3.5).

We call the first kind of system knowledge silo based, business area or subject area warehouses, and we refer to the second kind of system as holistic or integration warehouses.

3.2.1 Which type of warehouse is better, or which is the right one?

When we take a look at these different options for how we might approach the architecture for a warehousing environment, inevitably the following questions will come up for many people: Which approach is better? Which is the right one for me?

When we try to answer that question from a technological or business perspective, we will surely generate a wide range of conflicting attitudes and opinions. When we look at the question from a knowledge management perspective, however, we will see that all of the different types of warehouses have a critical role to play in the business.

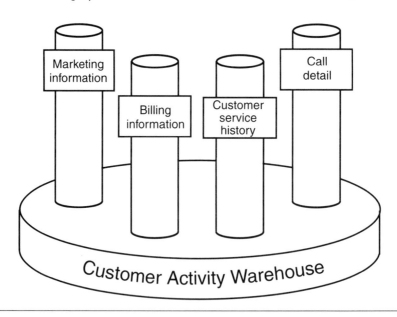

Figure 3.5 Customer activity, a holistic (cross-silo) warehouse.

3.2.1.1 The alignment problem

To understand why warehousing gets so complicated, all we have to do is go back to our original view of the organization and the alignment of computer systems with that structure. However, this time, let's consider the many interdependencies that these systems have with each other (see Figure 3.6).

Invariably, the organizational structure and the supporting information systems become unaligned over time. As the organization shifts and grows, and as the information systems are refined and optimized, we find a far less than perfect alignment between the two.

In fact, what happens very often is that certain key systems, often the most important system of all, becomes a critical part of all of the business units, with no clear alignment with any one group. For example, a company's billing system will often become a key part of the business' processing from all areas, but it will clearly not belong to any of them, as shown in Figure 3.7.

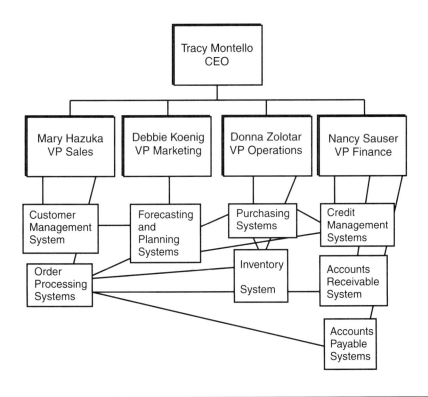

Figure 3.6 Interdependency of systems.

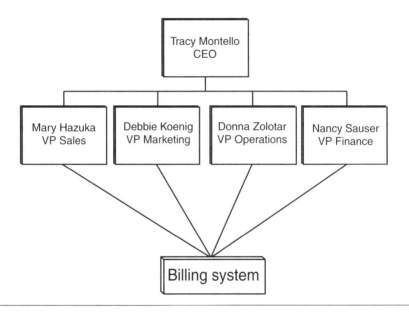

Figure 3.7 Common "shared" billing system.

3.2.1.2 How does this system interdependency happen?

The way this unalignment occurs is actually quite simple. It is part of the whole nature of the knowledge management problems that companies face.

When a business first establishes its systems, there is usually a very close alignment between the way the business is organized and the computer systems that support them. Over time, however, people begin to discover that the specialization and fragmentation that the business organization has created is starting to cause problems. The people in one area need to know what is going on in the other area. The response to this, usually, is to try to figure out a way to "merge" the two systems or to create a third one that will do the merging.

Eventually, as time goes on, more and more merging of systems occurs until finally you end up with so much interdependency and so many cross-functional systems that nobody is really in charge of any of them anymore.

3.2.1.3 Why is this merging of systems so bad?

At first, the solution of the company's fragmentation problems through the merging of operational systems makes perfect sense. When systems are new, they are relatively simple and straightforward. The merging of the systems adds a little complexity but, overall, it seems to be manageable. However, what few realize is that by taking this approach, you create two side effects that have devastating consequences in the long run.

First of all, as we try to merge and interface more and more systems, we begin to add astronomical increases in the overall complexity of those systems. Eventually, the systems become so interdependent that it becomes impossible to maintain or enhance any of them without creating problems for all of the other systems. The interdependencies get so complex and pervasive that we end up creating systems that require three months of work just to make a trivial change. The actual changing of the problem may be trivial, but the problem of figuring out what the consequences of that change may be on all of the other systems can be staggering.

The second reason this merging of systems is so bad is that when you try to take two highly efficient operational processes and "blend" them together for the common good, you are diluting the efficiency and effectiveness of each of them. You end up robbing Peter in order to pay Paul, as the saying goes.

The fact of the matter is that there is no "free lunch." Merging systems, while effective in solving some problems, actually creates several more along the way.

3.2.2 The warehouse alternative

It is at this point that we can begin to realize the real power and flexibility that warehouses have to offer. Warehouses make it possible for us to beat the information systems at their own game. With the warehousing option available to us, we no longer have to try to merge everything into one huge, megalithic system. We can use warehouses to optimize single processes (e.g., sales) or interdisciplinary ones.

3.2.3 A third alternative

There is a third way to view the construction of warehouses, and it is actually the best and most common way to look at them. That way is to build a warehouse in order to help solve specific business problems, as shown in Figure 3.8. These "special purpose" warehouses will pull data from whatever silo of information it needs to, and will often combine real-time and historical information to address whatever the need happens to be.

When placed all together, these options define a pretty comprehensive range of warehousing alternatives that need to be considered, and, of course, most warehousing activities involve a combination of some or all of these (see Figure 3.7).

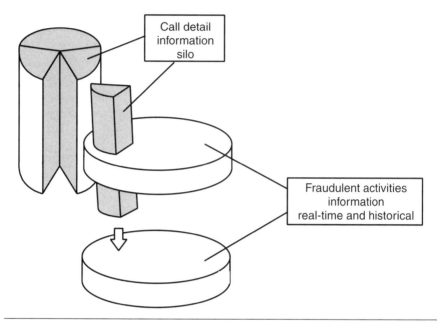

Figure 3.8 "Special purpose" warehouses.

3.3 The corporate global warehouse model

What these insights provide us with is a clear model of what the "perfect" warehouse would look like from a knowledge management perspective. What we would ultimately want to see is an environment that supports several "silos" of intense knowledge accessibility that cover each of the main areas of business knowledge and combine those with cross-functional integrating warehouses (from a real-time and historical perspective). This would allow people to better explore the synergies and efficiencies that the overall organization can obtain (see Figure 3.9).

Before going much further with our analysis, let's stop and take a couple of checkpoints.

First of all, understand that what we are mapping here is a knowledge model of the business, not a systems, organizational, or computer hardware model. The issues we must address before we are ready to turn this theoretical model into a working model are extensive. We perform this exercise only to provide ourselves with insight as to how to proceed.

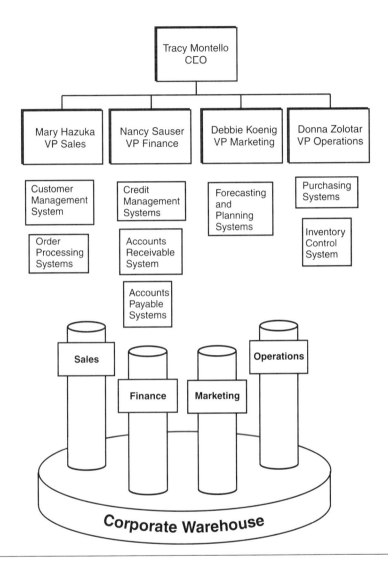

Figure 3.9 Warehouse architecture and organizational alignment.

Secondly, it is possible that the organization may never need or want to deploy an entire knowledge architecture along these lines. There are many organizational issues to be considered.

Thirdly, since this is a knowledge model and not a data model or system architecture model, we must realize that the stored computer data that we are talking about is only a very small piece of the sum total of knowledge that the company is storing. There are many other forms of computer data (documents, presentations, pictures, etc.), physical knowledge (books, magazines, reports, forms, etc.), and the knowledge in people's heads (the biggest source of knowledge of all) that are part of the knowledge landscape but are not part of our discussion here.

Lastly and most importantly, we only want to actually physically implement warehouse and mining activities in those areas of the business where this kind of knowledge is necessary or highly desirable. While it may be theoretically feasible to assume that a company with all of their information captured this way could be competitively superior to all others, the reality is that there are many organizational and physical implementation limits to consider.

While the idea of putting together a warehouse along these lines might seem like a good idea, we are still missing some very important information before we can make it usable. It involves the issue of what kinds of information we should put into the warehouse and how we figure out what has the most long-term value. In subsequent chapters we will be considering at great length the topic of what kind of information to put into the warehouse first and in the short term. In this chapter, we are concerned with the longer range consequences of how we organize it. The model we are developing will become our template for the future, a template that our short-term solutions can be laid into and structured, in order to gain us a combination of maximum short-term and long-term value.

To solve this problem of defining the long-term template for the warehouse, we need to use one more knowledge management-based tool, that being the concept of the corporate value chain.

3.3.1 Developing a truly usable global architecture model

It is easy to forget just how big, complicated, and unwieldy the modern business organization can be. Nevertheless, deciding just where the valuable information areas are should take first priority. Of course, every organization is different, and therefore the answer will be different each time.

Attempts to develop "architectural templates" for information systems have been tried many times before. There are several reasons why these have always failed.

First of all, these attempts to create the "perfect" templated industry-based model cannot take into account the idiosyncrasies of different organizations. An industry is a collection of similar companies that provide the same kind of products or services but who, in general, compete with each other. One of the biggest forms of differentiation that one company has is the way it organizes its people, processes, and information systems. Therefore, even if a good template did exist, companies would change their own model in order to gain a competitive advantage.

Secondly, the templates have tried to include both operational systems and warehouse (decision support) systems within the same model. This adds so many levels of complexity to the model that it becomes untenable in short order. The model we are proposing addresses the decision support/warehouse side only.

Lastly, these models are either based on purely theoretical grounds (subject area models), on processes (process-based model), on existing computer systems, or on organizational structure, all of which are subject to constant change. As businesses change and adjust to new technological innovations and to changes in their marketplace, their organizational structure, computer systems, processes, and theoretical foundations will shift, thereby invalidating the models in a relatively short amount of time. We will consider each of these alternative modeling approaches briefly in order to illustrate our point.

3.3.1.1 The logical modeling approach

Under the logical modeling approach to developing a system, we are told that all computer systems can ultimately be decomposed and organized into a collection of "logical" data collections. These logical collections are theoretically unchanging and constant. In order to design the "perfect" warehouse, therefore, we are told to identify the base collections and then build the rest of the system around them.

For example, we are told, there is a logical entity called "customer," and almost any corporate warehouse should have one well-defined Customer table from which all other systems will derive their customer information.

While this approach seems appealing on the surface, we create several problems for ourselves when we try this approach.

First of all, it is not possible to come up with a list of the core tables without getting into problems with information that belongs to more than one group. For example, consider the Customer, Sales, and Orders files. Clearly, these are three separate things, and clearly they are core entities for a business. But when it comes time to identify elements for each table, and the relationship between them, we get into all kinds of trouble. Where does customer name go? On the customer table, of course. What about the customer's billing address? Customer table also? Okay, what about the ship-to address? Now the question becomes a little less clear. What about the number of products the customer has ordered over the last three years? Well, now we think that might belong with

Sales or Orders instead of with Customers. The analytical process could go on forever. The fact of the matter that there is never a right answer to these kinds of questions, and so we play logical modeling ring around the rosy for the rest of the project's life.

3.3.1.2 The process-based model

Another way to try to derive the best warehouse model is to attempt to base it on the critical business processes: billing, sales, ordering, purchasing, and so on. Each of these major functions should have their own store of data. But just as in the case of the logical models, we find too many situations where there are conflicting demands upon the information, and we quickly get caught up in the same kind of analysis paralysis.

Along with that problem is the predicament caused by alignment issues. Which group of knowledge workers "owns" the information? While some collections may clearly belong to one group or the other, other collections will undoubtedly belong to everyone.

3.3.1.3 System- and organizational-based models

Many companies have tried to build warehouses based on existing operational system-based or organizational-based models.

For example, for every major operational system (billing, ordering, production control, etc.), we create a warehouse to store the information contained therein. Actually, of all of the options we have discussed, this approach has the most merit. At least the problems of ownership, architecture, and acquisition are greatly simplified (see Figure 3.10).

Unfortunately, operational systems are subject to change in far too many cases, and building our model upon that shaky ground is asking for trouble in the long run.

On the other hand, it goes without saying that a warehouse designed along organizational lines (one for marketing, one for sales, etc.) also has some merit. But organizational structures tend to change fairly regularly in business these days, making it an equally weak foundation.

3.3.2 An alternative foundation: the value chain

What we have found to be most effective in creating some guidelines for making these decisions is to put together what is known as a value chain analysis of the business.

The point of a value chain analysis is to attempt to cut through the "clutter" created by the organizational and political structure and figure out what the core

The knowledge management view of business and warehousing 43

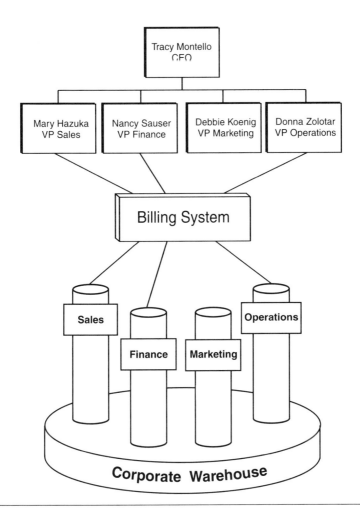

Figure 3.10 The "alignment" problem.

competencies and objectives of the business really are. When you do a value chain analysis, you ask yourself some simple and very direct questions. What is the point of our business? Who are our customers, and what needs do we meet? And then, finally, which are the core business functional areas that drive the process of meeting those needs? When we get the answer to this last question, we identify the business' core value chain.

The challenge to doing an analysis like this comes when you evaluate the existing organization (and the subsequent computer system's footprint created by that organizational structure) and try to figure out what is critical and what isn't. It is actually very industry dependent and business specific.

For example, in a services company like a law firm or a consulting business, human resources systems are going to be the key information repositories that drive the business. In the consulting business, human resources are the product.

On the other hand, in the coal mining business, core business value drives from people (human resources), equipment, property, and processing facilities.

In the insurance industry, human resources become even less important as marketing, underwriting, actuarial, and sales are the key areas, as shown in Figure 3.11.

3.3.2.1 Walking through the value chain

The first thing we do when we develop a value chain is figure out what the point of the business is. In the insurance industry case, it is to provide security to customers through the issuance of insurance policies. Notice, the main "driver system" for any insurance company will always be policies. (In every industry there will inevitably be one or a series of core driver systems upon which all other systems and the entire business are based.)

Basically, the creation and maintenance of policies is the point of this business. So in this case, the value chain is defined by figuring out which major functions need to be performed in order to create, package, and sell those policies, and then pay out to customers when they need it.

The main processes in the value chain are, therefore, marketing (figuring out what kinds of policies to sell), actuarial (how to price the policies), underwriting (crafting the specific policy solutions for customers), processing claims (paying out to clients), and sales (finding customers to sell policies to).

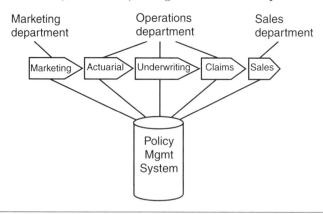

Figure 3.11 Value chain example for the insurance industry.

We can then see listed above the value chain the different "departments" that manage each of the functions in the chain.

Our proposal is that with the value chain we have identified for the business, you can create a template for the development of a warehouse that will stand the test of time. The organization may change, and the operational systems may change, but the core process that it takes to deliver the goods to the customer (and the subsequent types of knowledge that must be managed in order to do so efficiently) will remain stable.

3.3.3 The key to value chain delivery

At this point it is important for us to understand what the consequences of our analysis are showing us. We have seen that there are organizational structures, and business functions and processes that do work in order to meet customers' needs, and operational systems that align with those functions and processes and make it possible to deliver them. We have also seen that there is a shift in the emphasis of the business away from managing the processes (businesses have mastered that in many ways) and towards the management of the knowledge that the business holds.

This makes three things clear:

- Operational systems are about optimizing the management of processes throughout the organization.
- Data warehouses and data mining are about optimizing the management of knowledge throughout the organization, but these systems can only be fed by information that the operational systems hold and manage.
- The organizational structure is about the management of people, to manage them to levels of optimum effectiveness and efficiency.

Therefore, our optimum data warehousing environment can only be developed when we understand the relationships between all three of these aspects of the business and we can figure out how to align them effectively.

3.4 Overall strategy for development (one piece at a time, fitting into the overall architecture)

The job of figuring out the complete and perfect alignment between all of these different aspects of the business is clearly going to be a large, complicated, and arduous task. The knee-jerk reaction of the theorist at this point will be that the company should commission a study group or hire a consulting firm to come in and figure all of this stuff out.

For many reasons, this is probably not a good idea. Mostly, it will not be possible to do this in any reasonable amount of time or at a reasonable cost. The issues that will be uncovered when trying to figure out all of the interdependencies and relationships will be overwhelming physically, logically, and theoretically. The danger will be that we could very quickly become "lost in the weeds" and spend an inordinate amount of time trying to polish the model instead of worrying about the business of servicing customer needs.

For that reason, we do not propose here to go about the process of developing your warehouse in this way.

You may remember that we identified three primary ways to look at the warehousing process from the knowledge management perspective. We can look at the cross-silo integration capabilities, the optimization of silo activities view, or we can look to the point in time, special purpose warehouse initiatives designed to solve immediate, pressing business problems.

Clearly, for any number of reasons, the place where we want to start is in the area of deploying effective, short-term solutions quickly. In fact, it is in the execution of these types of projects that data warehousing and data mining have shown the best success. The telecommunications industry and indeed all industries are rife with case studies of organizations that deployed warehousing to solve a particular problem and found that not only were their immediate problems solved, but that several other unanticipated benefits were discovered along the way.

In fact, as a rule of thumb, we say that any warehousing effort you pursue should

- Be operational in less than six months;
- Solve immediate tactical business problems;
- Be cost justified and, in fact, deliver at least a five times return on investment within the first 12 months of its life.

Now, that may sound like a tall order when you first look at it, but the reality is that companies are delivering that kind of return on investment within those kinds of time frames on a regular basis. During subsequent chapters we will consider in much more detail exactly how this is done, and how you can do it too.

But the more pressing issue at this juncture is this. While it is possible to deploy small, tactical warehousing solutions in order to solve immediate business problems and achieve substantial returns on investment, how do we go about organizing that process so that five years from now we do not find ourselves having to deal with dozens or even hundreds of mini-data warehouses, all of them delivering value to the business but creating yet another generation of highly interdependent, severely intransigent, and incredibly cumbersome isolated pockets of information, which will make it even harder for the company to change in the future.

Our answer to this problem is to use our earlier discussions about value-chain-driven architecture as the key.

While it is highly recommended that you limit your warehousing efforts to those areas where the business needs them the most, we also suggest that when you begin deploying those systems, you have a much bigger, enterprise-wide vision in mind. This vision should be anchored in a logical and physical architecture that is derived from your own understanding of the organizational structure and the subsequent value chain that drives the businesses survival.

In other words, you should start by imagining an environment where every major silo of information (value chain element) represents a core warehousing area. Within each of these areas will reside all information pertinent to that group of knowledge workers (Sales, Marketing, Finance, etc.). At the same time, you should have constructed a neutral, globally defined cross-silo corporate warehousing area that will house all information that is shared, merged, or otherwise transported between those silos, as shown in Figure 3.12.

When you then set out to deploy specific tactical solutions, you can begin to overlay them into your architectural model. An example should help illustrate how this process can work.

3.4.1 The growing warehouse example

Assume an insurance company, like the one we have been discussing, where we have already figured out what the organizational structure, value chain, major processes, and operational systems look like.

The first thing we want to do is identify those places where the warehouse can deliver immediate value. In this case, we'll say that Marketing has proven that if you could provide them with a marketing database, holding customer buying behavior and demographic characteristics, it would be possible to prequalify new customers and save the company many millions of dollars a year in prospecting costs. Clearly, this should be the first project, and so you build a miniwarehouse to meet that need.

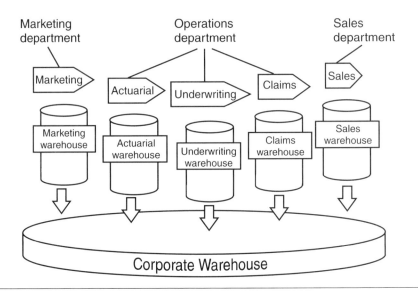

Figure 3.12 The global environment.

The second thing that happens is that the Claims Processing group discovers that they can reduce claims for traffic accidents by gathering a database full of past claimant information and use data mining tools to detect the patterns of current customers who get into accident situations and to help those people avoid those types of situations, thereby reducing claims. Our second miniwarehouse is born.

With the information about customers in the marketing miniwarehouse and the information about claims available in the claims miniwarehouse, suddenly someone realizes that by combining the information from both systems you could actually figure out how likely prospective customers are to be involved in accidents and use that to assist in the underwriting and actuarial areas.

In this case, we realize that since the information is cross-silo in scope, the new system should be built in the "global" area, thereby building up the inventory of the cross-silo warehouse as well.

As time goes on, we continue to use the same decision-making criteria:

1. Build only those parts of the system that provide immediate value.
2. At the same time, place those miniwarehouses within the architectural layout where they best fit.

Eventually, you could very well end up with a complete global- and silo-rich warehousing environment without ever having to stop and figure out all of the detail up front.

Of course, there are many more issues for us to consider before you go running off to try to develop a solution along the lines that we have described, but the general model for how this works is valid (see Figure 3.13).

3.4.2 Ownership of knowledge issues

While we will be using the majority of the rest of this book to help flesh out the details of how this kind of warehousing template can be developed and deployed for a telecommunications company, there is one more generalized issue that we want to discuss, and that has to do with the organizational structure and its alignment with the warehouse architecture.

Up until now we actually have not developed a very strong case for allowing our value chain and our subsequent architectural template to be driven from the organizational chart. Why is that so important? Couldn't we, in theory, simply construct the warehouse based on what logically flows from the value chain analysis? Isn't that in itself enough guidance for us? The answer, unfortunately, is an unequivocal no.

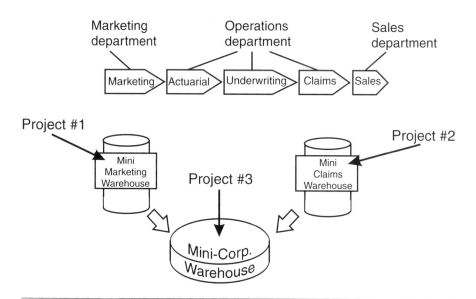

Figure 3.13 Stepwise construction of the global warehouse.

In actuality, making sure that the organization and the warehouse line up is probably the most important key to making it successful, and the reason is pretty obvious when you stop and think about it.

We can theorize, conjecture, and wax philosophic all day and all night about the benefits of data warehousing and the different ways it can help. We can even build strong cases that rationalize building different parts of the system or even the overall warehouse structure.

But, unfortunately, nothing gets done in the business world unless there is a business manager—responsible for an operational area of the business, with a staff and a budget and bottom-line revenue and expense responsibilities—who is willing to commit the funding for the project. And only the most altruistic, intelligent, and farsighted business manager in the world is going to fund a multimillion dollar data processing project that doesn't deliver immediate and substantial value to his or her particular area of the business. While this may seem selfish, shortsighted, and not fair, it is a fact of life.

So, what this means is that we will need to put our architecture together based not only on what the logical and functional realities of the business are, but on the budgetary and political realities as well. No manager wants to pay to help some other manager.

Therefore, we will inevitably be forced to build just as many different miniwarehouses as it is going to take in order to get the job done.

The good news is that over time, as our global warehouse becomes more "filled in," our dependence on these conditions will become less and less.

In the next chapter we will begin to apply this theory to practice by taking a look at telecommunications organizational structures, processes, and typical value chains, and with these propose some candidate warehouse templates that you can use as starting points for your own analysis.

Chapter 4

The telecommunications value chain

So far we have established several things about the process of data warehouse development, many of which are not so obvious at first glance.

We have established that data warehousing and data mining technologies can deliver value to the telecommunications firm on a problem by problem basis. In fact, we have cited many cases where telecommunications firms have done exactly that.

One approach might be, given that information, to proceed full speed ahead toward the identification and deployment of warehouses to solve specific problems in your own organization. Unfortunately, it is painfully true that if we do not come up with some kind of an overall organizational approach to the process, several negative things will happen.

First, computer systems and planning personnel will begin arguing and strategizing about what the "best" physical and logical architecture for this system should be. Making these decisions will be important in the long run. However, trying to develop the answers to those kinds of questions could take years, and no one can wait that long to get something done.

Second, even if the physical infrastructure was not going to be a problem, proceeding without some kind of a long-term plan could lead us into a situation where we create dozens or even hundreds of little "data marts" and point solutions that will turn our future environment into a data, information, and knowledge management nightmare.

Third, as an environment like this develops, there will surely arise situations where it will be possible to leverage work and activities performed in the creation of earlier warehouses, which later ones can use to their advantage, greatly reducing the cost of the new solutions. In fact, this kind of leveraging is one of the big attractions for warehousing as an overall approach.

Ultimately, we need to develop some kind of a long-term framework into which our short-term, immediate-business-value-type applications can fit.

4.1 The knowledge roadmap solution

We then proposed that the best way for putting together this kind of long-term roadmap would be to base the approach upon the concepts of knowledge management (viewing the system as an extension of the knowledge management capabilities of the firm). The driving icon for this effort should be the company's unique "value chain," which is nothing more than a diagram that illustrates the key business processes that apply knowledge directly to the delivery of value to the end customer.

We finally proposed that the only way to put this kind of analysis together would be through the development of a thorough understanding and subsequent alignment of the organizational structure, the computer systems environment, the business functions being executed and, finally, the knowledge-based value chain itself.

4.2 Steps in the process of deriving a business' value chain

In order to assemble a value chain for any organization, we need to begin by addressing the many different kinds of alignment issues we have been discussing. At this point, it probably makes sense to spend a moment on this issue of alignment. In the previous chapter we showed how the company's organizational structure and information systems tended to cluster together into loosely related knowledge areas. We also showed that when you start to examine those systems more closely, you begin to discover a lot of inconsistencies in that alignment.

While on the one hand we are insisting that it is imperative that we discover and understand exactly where these misalignments between organization, function, computer system, and knowledge needs might be, on the other hand we in no way imply or insist that some kind of fully logical or comprehensive strategy needs to be put in place in order for us to proceed. On the contrary, we perform our analysis of this alignment to help us understand where our warehousing efforts may provide value by filling in the gaps between organizational groups or functions. Although some overambitious and enthusiastic well-meaning theoreticians might try to take this exercise and turn it into some kind of all-encompassing theory of information systems architecture, we state emphatically that our only purpose in doing so is to provide us with a generalized roadmap of the directions that our warehousing efforts should take us.

4.3 Telecommunications functions and systems

We will begin our search for the telecommunication firm's typical value chain by getting a good understanding of just what the basic functions are that a telecommunications firm must exercise in order to deliver value to the customer. As we consider each of these functions, we will also take a brief inventory of the nature of the computer information systems associated with those functions. This will provide us with our first level of understanding of the telecommunications company's knowledge infrastructure and potential knowledge management needs.

Since we include in our discussions a wide range of segments (traditional telephone, long line, cellular, PCS, cable/developmental, CAP, etc.) within the telecommunications industry, we will borrow terminology and concepts from each of them to present an all-inclusive impression of a "typical" firm in the industry. Obviously, for each individual company of a different size, shape, history, culture, and industry segment, the analysis will vary, but our model should remain relatively applicable across the board.

4.3.1 Creation (new product development and exploitation)

For those telecommunications organizations that are later in their life cycle (long lines or traditional services), it is difficult to remember those early days of telecommunications when the concept of "a telephone in every home," let alone the "a telephone in every hand" concept of today, was new, radical, and socially inconceivable. So, for those kinds of companies the creation process is pretty much taken for granted and assumed to be *done*. However, for a large percentage of the telecommunications industry, especially the PCS, cable/tele and other high-creativity, high-risk entrepreneurs, the lion's share of the company's investments are going into the identification and creation of new kinds of products and services to meet customers needs.

The information needs of the creation function are usually large and include the need for marketing information, customer activity history, and consumer trends analysis, along with the inclusion and dissection of vast amounts of technical information as alternative "product offerings" are tested and improved.

While the role of creation will take on a different emphasis depending on the company, it certainly needs to be included in our "generic" model.

As far as information systems deployment goes, creation needs to be one of the biggest *takers* of knowledge from other areas of the business, and also needs to be one of the smallest *givers*. Creation activities generally happen without the benefit of almost any "operational" systems of any kind (except of course in the cases of *monster* R&D companies like AT&T), and usually provide no output to other areas of the firm until a product is "created," at which time the entire organization is restructured.

4.3.2 Acquisition (acquiring the "right" to do business)

The vast majority of "real" telecommunications companies today may invest some small portion of their budgets in the creation process, but for the most part they leave those kinds of efforts to the small, startup "pioneer" or R&D firms that function separately from the main telecommunications providers.

The one thing that every telecommunications company *must* do, however, before anything else, is to acquire the right to service certain geographical or legal areas and work to maintain that right.

4.3.2.1 Acquiring the "right" to do business

The nature of the right to do business and the way it is acquired vary with the country, the industry segment (cellular, PCS, long line, traditional service, etc.), and the economic sector (government, industry, residential, commercial, military, private, public, etc.), but without fail someone has to give you permission and the right to service their area of concern.

The acquisition and maintenance of the right to service an area may require no more knowledge or experience than supporting a legal staff that resolves boundary disputes and protects territorial claims. Or it could be as all encompassing and exciting as the process of bidding on huge blocks of market share/radio frequencies being auctioned off to the lowest bidder by government agencies.

In all cases, the acquisition and maintenance of the right to do business is a foundation of the telecommunications company's existence. In the "olden days" of telecommunications, this right was usually taken for granted. Old-form telecommunications companies in the United States and in most countries around the world were established as a government-sanctioned monopoly or a division of the government itself. Today, you not only have to "win" these kinds of rights, you often have to fight to keep them, and you *always* have to report to the agency that granted you those rights in order to prove that you are worthy of keeping them.

Just as in the case of the creation activity, the acquisition function provides very little input to the rest of the business in terms of traditionally based computer data, but it does create vast quantities of nontraditional documentation, maps, presentations, and other forms of knowledge transfer, which form the basis for the entire telecommunications organization.

The acquisition function also requires that the rest of the organization provide a substantial amount of reporting-type information in order to support the legislated and mandated requirements of various government committees, departments, agencies, and other administrative organizations. These reporting requirements can at times become quite substantial. Traditionally, the acquisition area of the telecommunications firm will have no computer systems of its own, but will frequently tap into the information systems of other groups to gather the information required.

4.3.3 Network infrastructure planning and development (creating the "phone system")

Assuming that the company has a known product to sell (e.g., traditional phone service) and has acquired the right to provide that service to a specific area, we come to that point of the business where everyone has to start producing—that is in the area of network infrastructure development.

The people involved in this process are the ones who decide where the cables need to be laid, where the switching stations will be located, what equipment will be placed into the stations, and in general create what is known publicly as "the phone system."

Now, obviously, depending on the technology you are talking about and the age and history of the particular company, the knowledge management demands of these groups can be varied. But by the nature of the business you will have to know about and plan for certain things.

This area is also responsible for the identification, understanding, and engineering of cross-carrier interfaces, that is, making it possible to "hand off" traffic to long-line carriers, CAPs, and the other "partners" necessary to create a complete communications service to the customer.

The key data requirements for the engineers in this area have to do with developing estimates for the volume, nature, and intensity of different types of traffic that the system is expected to carry. These specialists develop estimates of that activity and then use those estimates to determine how best to deploy different throughput and bandwidth traffic scenarios.

The areas of the business concerned with these kinds of activities are usually inundated with several different kinds of specialized hardware and software, depending on which aspects of the system you are talking about. This area determines what kinds of switching mechanisms will be deployed and how that information will be made available for other areas of the business.

This same area will also require information about historical calling patterns (from the switches); geographical, topological, and a wide range of scientific and meteorological data (to help in deciding the how and where); billing and customer activity levels (from the billing area); and projections of increased demand requirements (based on information gleaned from outside sources or from marketing).

4.3.4 Network infrastructure maintenance (maintaining the "phone system")

Given that the network infrastructure is up and running, the next major area of the business to consider is the network maintenance group. These are the people who will keep the network running throughout the lifetime of the system.

Network maintenance people typically work with a wide range of computer systems. Customer service, customer trouble call management, work ticket management, and a wide variety of similar types of systems are the most common. These departments are the arms and legs of the telecommunications firm.

This group responds to three different causes for activity:

1. A report of system failure;
2. A request for system alteration or enhancement;
3. The installation of new service (in the traditional telecommunications or cable case).

Usually, trouble calls are reported to some kind of customer service personnel who then inputs the information into a system that notifies network maintenance and monitors the process.

This area, too, needs information from other sources within the business. Marketing, sales, customer service, network development, and creation information are all necessary for the network maintenance people to do their job.

4.3.5 Provisioning (setting up customer services)

One of the most frequently executed areas of activity in telecommunications is the provisioning process. To make it easy for the company to "activate" customers as quickly as possible, more firms "preactivate" a large block of numbers, making them available for relatively immediate activation when a customer asks for service.

Activities in this arena involve registering numbers with the line identification database (LIDB, the master list of all phone numbers in the world) and presetting the switching and other network monitoring systems to recognize and handle the pre-identified numbers appropriately.

The provisioning areas' information needs are usually driven by the type of telecommunications firm, the traditional method for handling the process, and the lead times required for the delivery of full service to the customer. In this and in the activation areas we have been seeing a lot of automation over the past several years. These provisioning systems must be tied off to the billing, switching, and external provider (LIDB) interfaces if they are going to be effective. In those companies where the process is not fully automated, manual forms carry the provisioning information to the appropriate areas where it is then keyed into their systems.

4.3.6 Activation (activating customer services)

Provisioning is the process of "priming" the system for a particular customer, and activation is the process of bringing that individual customer's service alive. Again, the nature of the activation process varies, sometimes dramatically.

In the cellular area, activation is particularly challenging because the typical cellular telephone on the market today can be used to support many phone services. As a consequence, "activation" consists of programming the communications chip within that unit to recognize the phone number the selected carrier decides to assign. Just a few short years ago, this activation process could only be done at specialized "activation sites," making the process of getting a cellular phone a time-consuming one. Recently, onsite, easy to use, remote "activators" have removed this obstacle.

Unfortunately, with the new ease of use capability came a new problem: cloning. The very capability that relieved a major stumbling block to the deployment of cellular phones turned around and became one of its biggest curses.

Activation, like provisioning (sometimes the terms are used interchangeably within certain firms or industries), may have a large automated computer system driving the process, or it may be a process handled by forms and people.

4.3.7 Service order processing

For many of the larger telecommunications firms, the processes of activation, provisioning, and network maintenance become such a large and complicated set of tasks that a special form, called the "service order," is created. The service order becomes the standardized form that documents and drives the process of getting the work done.

In those cases, service order processing becomes a kind of superset process and encompasses all of those other functions under its umbrella. In many large firms, separate service order processing and management systems are created to help in the complex tracking and monitoring of these activities.

4.3.8 Billing (tracking service and invoicing the customer)

All of the business functions that we have been talking about up to this point happen immediately around and are focused on the development and maintenance of the network side of the telecommunications business. These functions represent only half of what the business is about. The other half occurs on what we will call the "business" side of the house (as opposed to the "network" side).

The first and most critical function in any telecommunications company, short of providing the actual service to the customer, is the creation and execution of customer bills. While the billing function may be managed by different departments within different organizational structures (Customer Service, Operations, Billing, or a host of others), it is always the heart of the business side of the telecommunications company.

The billing system is inevitably connected more or less directly to the switching and control system (on the network side) and becomes the system of record for the tracking of customer activity.

4.3.9 Marketing (identifying prospects/channels, advertising)

While in the "good old days" of telecommunications there was no need for a real marketing function of any kind (since the customer population was forced to do business with whoever the legally sanctioned provider happened to be), in today's competitive environment telecommunications firms are finding an increased dependence on the presence of a good, strong, well-integrated marketing function.

In terms of the importance of marketing within the telecommunications concern, it depends on the industry segment (cellular, long lines, etc.) and the competitive environment. Some of the older, more traditional telecommunications firms maintain a relatively small "advertising"-based marketing posture, while others are investing heavily in large marketing databases and other forms of consumer tracking and forecasting systems.

In all cases, as we noted in the first chapter, marketing is playing an ever increasingly important role. In the strongest cases, marketing not only becomes the driver of customer identification and pursuit, but takes a commanding lead in the issues of channel development and management (finding alternative ways to sell your service to consumers), the product creation process (taking the lead in defining what the new product offerings will be and how they will be positioned), and acquisition (providing leadership in the determination of what the potential market value of different geographical licensing agreements might be, thereby directing which licenses to pursue and which to ignore).

There is *no* telecommunications company in existence that does not have at the heart of its operation a large, complex billing system, and this system usually serves as the main source of information for all other "business side" processes and people.

4.3.10 Customer service (keeping the customer happy)

Another key area of the telecommunications firm can be found in the establishment of the customer service function. Customer service involves all of those jobs that keep the company in direct contact with the customers. Customer complaints, requests for information, requests for changes in service, resolution of billing problems, and many other services fall into this area.

Like marketing and sales, customer service in telecommunications companies is undergoing some significant revamping based on the new way companies are looking at their markets. While the customer service area has always been strong, it has been changing from being a very large, cumbersome, and "passive" type of organization (where agents sit waiting for customer problems to arise), to becoming a proactive part of an overall marketing strategy.

Customer service departments are now taking on the additional responsibilities of functioning as outbound phone canvassing and phone solicitation agencies (calling out in attempts to locate new customers), while the inbound call handlers are being asked to more immediately research customer complaints, resolve them quickly, and try to use the opportunity to sell more of different kinds of services to existing customers.

The original customer service computer systems were mostly deployed as combination bill tracking and resolution systems and/or customer service order creation and tracking systems.

These antiquated and rudimentarily functioning systems are being replaced by multifunctional, fully integrated phone room automated call processing centers.

Inbound calls from customers are handled by computers and computer-generated voice attendants who determine what the customer wants and who to route the call to. Many of the more rudimentary customer requests are handled completely by the computer itself. For example, you can check on your account balance or on the status of your request for new services by simply pressing the right keys on your touchtone phone.

Calls that require human intervention are routed to the correct specialist. In the most sophisticated systems, the person answering the phone is presented with a screen full of information about who the customer is, what their billing and service history is, and what kinds of problems they have had in the past before they ever pick of the phone.

On the outbound side, automated dial-out services and integrated database marketing facilities generate lists of people that the telemarketing person is supposed to call. This person doesn't ever actually pick up the phone and dial, however. Instead, the system itself begins dialing all of the desired numbers until it finds a phone that somebody answers.

The person who actually answered the phone is routed to the next available telemarketing specialist. This specialist does not then have to try to figure out what to say to the person. In this case, the system will have checked on the purchasing and demographic characteristics of the person they will be talking to, and will automatically post a predefined script onto the screen. The script not only instructs the telemarketer of what to ask the prospective customer, but it actually dynamically provides the response to be given to whatever the prospect has said.

These powerful telemarketing and database marketing techniques and facilities are quickly turning customer service into one of the most important growth areas for the telecommunications firm, and many firms are responding through extremely heavy investment into these areas.

4.3.11 Sales (establishing and maintaining customer relationships)

Again, depending on the structure, size, and segment of the telecommunications firm, the sales function can be an important driving force, carrying the company into new markets and deeper penetration of those markets, or it can be nonexistent, with other functions (marketing, customer service, etc.) doing the same tasks.

In those cases where the sales function is formalized, there are usually few information systems to support them and even less information made available to them. Again, new generations of automated sales support software are being deployed that tie the sales, marketing, and customer service functions tightly together, helping them to work in a coordinated fashion.

4.3.12 Finance and accounting

Of course, no discussion of business functions or processes would be complete without the inclusion of finance. Finance is one of the three major areas (human resources and information systems being the other two) that traverses all business functions and attempts to help glue them together into a cohesive whole.

No area of the telecommunications firm can function without the input and support of finance. Therefore, in terms of knowledge management, we find finance as an important *support* function, but *not* a critical deliverer of value to the customer function (as opposed to the banking industry, for instance, where finance *is* the service being provided to the customer).

Although not part of the direct telecommunications company's value chain, finance is the source, keeper, and disperser of much of the information and knowledge critical to the success of each of the business areas. Finance systems will be a key component of any warehouse architectural strategy.

4.3.13 Credit management

While in some organizations the credit department is just another part of the accounting function, in some companies (like cellular and PCS) it becomes such a major component of a successful operation that we need to consider this function as a critical part of the value chain in and of itself.

When the decision of where, when, how, and why to grant customers credit for the use of your phone system becomes a critical part of the marketing strategy, then keeping track of customers and the credit risk profile becomes key.

4.3.14 Operations (network and business)

In many companies, organizational units called *operations, network operations,* or *business operations* will be prevalent. These organizations, while important, are usually umbrella groups that include within them some collection of the functions we have already described. For example, for one telecommunications firm, operations will refer to those people involved in the network infrastructure development and maintenance areas. For another, it will refer to a combination of customer service, billing, sales, and marketing. Use of the term is company specific.

4.3.15 A comprehensive value chain

Assuming the absolutely perfect and all-encompassing telecommunications company, we might come up with an initial value chain that looks something like Figure 4.1. As we can see, each of the functions (creation, acquisition, network development, network maintenance, provisioning, activation, service order processing, billing, customer service, marketing, sales, and finance) claims a role within the process of delivering value to the end customer.

In reality, of course, the actual value chain we want to build will be simplified and customized to the individual company to make it truly useful.

At the very least, what we will discover is that no telecommunications firm ever has everything from the ultimate value chain as part of their analysis. Some firms, the new startup firms for example, will find themselves heavily invested in the creation and acquisition parts of the process, with very little need for information from the later stages. A more mature company, in a more mature segment of the industry, may find that creation and acquisition are relatively trivial issues. Their products have been created and acquired already. Theirs is the business of capitalizing on them.

Not only will different companies make use of only selected parts of the value chain, but they will place a different emphasis and a different degree of importance on each. Some companies will find that concentrating on marketing is key, while others will emphasize operational efficiency. In fact, the choice to

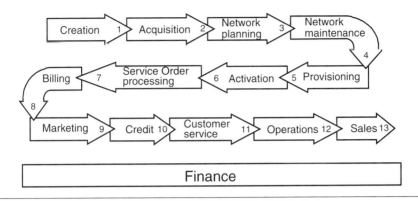

Figure 4.1 The ultimate telecommunications value chain.

emphasize different parts of the value addition process is precisely how companies try to differentiate themselves from their competitors.

Finally, we also run into the problem of having to deal with telecommunications companies that are extremely large and diverse, and therefore find themselves representing many different companies and products, all beneath the same umbrella. In this case, we the company will be looking at several value chains, one for each line of business.

Our next challenge becomes, therefore, figuring out what the actual value chain looks like for a given company (or segment of the company). How can we possibly weed through all of the different characteristics of the business and come up with a value chain that makes sense? Well, to do this we have to turn, first and foremost, to gaining an understanding of how the organizational structure and the value chain relate.

4.4 Organizational structure and the value chain

Given our basic understanding of the functions involved in the deployment of telecommunications services to customers, the next question we need to ask is, "How does the individual company organize itself around the chain?" It is here where we can start to get very confused very quickly.

For, while every company must somehow perform each of these functions, no company can get all of these things done in institutionalized (as reflected by the org chart) or even formalized (as reflected in stated standards, policies, and procedures) ways.

Each company places different emphasis on different parts of the value chain depending on the nature of their business, the nature of their market, their maturity as an industry segment and as a company, and on their physical infrastructure, consumer reputation, and information systems legacy.

4.4.1 Typical organizational structure: medium-sized cellular firm

A typical medium-sized cellular firm will have an organizational structure in many ways similar to other kinds of business. The company will be headed up by a CEO/president who will have several vice presidents reporting to him or her. Each of the vice presidents will be the head of a major area of the business. Typically, these will include operations, marketing, information systems, finance, and engineering (see Figure 4.2).

While this org chart tells us how different groups of specialists in the firm have been organized, it actually says very little about which value-added functions each of the groups provides. (For example, who is responsible for customer service, sales, and other customer contacts? It could be marketing, operations, or even finance. Who handles the billing functions?)

Therefore, in order to make the organizational chart meaningful for our purposes, we need to overlay it with our ultimate telecommunications value chain in order to understand how things really work.

In Table 4.1 we show which value adding services each part of the business is responsible for.

Table 4.1 Business Area Responsibilities

Major Area	Value Chain Coverage
Operations	Customer service, billing, sales, activation
Marketing	Creation, acquisition, marketing
Engineering	Network development and maintenance, provisioning, activation
Finance	Across all areas and processes
Information Systems	Across all non-"network"-based computer systems

The telecommunications value chain 65

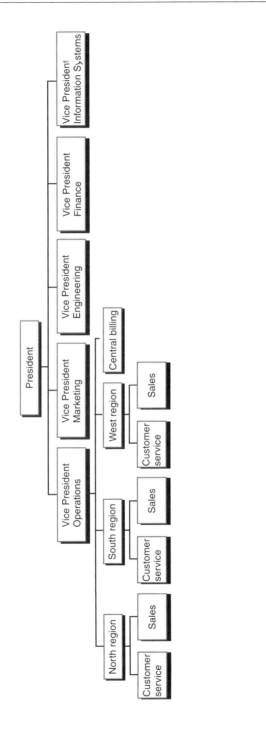

Figure 4.2 Typical medium-sized corporate org chart.

When we can create a chart like this for an organization, we can say that we have completed our organizational/value chain alignment analysis.

Notice how some areas of the business have responsibilities across the breadth of all or many of the value chain components [i.e., finance, information technology (I/T)] while others have full responsibility for several of them, and still others are shared by more than one (i.e., activation).

Obviously, each of these major business areas can be decomposed into their smaller operational units, and if we were to do this kind of breakdown for this particular company, we would find that there are similar kinds of breakouts of alignment down to the lowest level. Ultimately, we would find that every employee and every process is in some way either directly or indirectly involved in the course of delivering value to the customer. Of course, what we are most interested in are those people who have direct, immediate, and substantial impact on that process.

While the exercise of aligning the organizational structure to the value chain proved to be a relatively simple process in the case of the medium-sized telecommunications corporation, how do we go about analyzing the larger types of firms?

4.4.2 Typical organizational structure: large telecommunications firms

A typical large American traditional consumer and business phone service provider, having millions of customers and thousands of employees, will often be organized along very bureaucratically complex and verbose lines. Many phone companies issue full phone books with the names and office phone numbers of all their employees.

Table 4.2 shows one such phone book that is made up of a collection of entries at the highest level.

4.5 Allocating the business units to the value chain and the knowledge management process

Even at this very high level, for an organization of this size, we can begin to see a little of how the organizational structure breaks out against our knowledge management framework. For example, of the areas listed, the network services business unit clearly corresponds with our network development and

Table 4.2 Major Business Units

Major Business Unit Name	Value Chain Coverage	Market Segment	Profit/Support Unit
Cellular Services	Complete value chain	Self-contained	Profit center
Consumer Services	Product creation, marketing, customer service	Overlaps with profit centers	Support unit
Information Services	Product creation, marketing, customer service	Overlaps with profit centers	Support unit
Leasing Services	Product creation, marketing, customer service, finance	Overlaps with profit centers	Support unit
Long-Distance Services	Complete value chain	Self contained	Profit center
Network Services	Network development and maintenance	Provides same support for all profit centers	Support unit
Pay Phone Services	Product creation, network maintenance, marketing, customer service	Overlaps with profit centers	Support unit
State Organization (Geographical Region) #1	Billing	Customer service, sales, marketing, finance	Overlaps with support units
State Organization (Geographical Region) #2	Billing, customer service	Sales	Marketing

maintenance descriptions. In this case, the business unit and the value chain position are in alignment.

The vast majority of the other major business units in this example represent a variety of special kinds of organizational structures called market segment-based specializations. In these cases, the company has determined that a certain segment of the marketplace requires special attention, and an entire organization is built around making sure that the appropriate resources are focused in those directions.

There are several different ways that a market can be segmented and, consequently, covered through the creation of special business units. Some of the more popular approaches include:

- *Geographical models*—As in the case of the different state organizations in our example above, in this case each business unit is given responsibility for managing all consumers within a certain geographical area.
- *Consumer category models*—Corporate, consumer, pay phone, and so forth. In these cases, separate business units develop specialized products, services, and marketing campaigns to meet the needs of these specific areas.
- *Line-of-business based*—Cellular, long distance. These business units are created to cover the product lines that are so different from the main business that they cannot be supported by the existing infrastructure.

Not only does the creation of these "overlapping" or matrixed organizational structures create plenty of opportunities for discontinuity in the efficiency of the delivery of services to the consumer, but the problem can be made even more complicated by the allocation of ultimate revenue responsibility and designation as a "profit center" or a "support unit."

Profit center business units have ultimate responsibility for the relationship with the customer and for collecting the revenues earned. Support units tend to create specialized programs and offerings, but rely upon the profit centers for the actual delivery of the services.

In Table 4.2, you can see our list of major business units and the way they break out according to these different criteria.

4.5.1 Aligning the value chain and the organization—large megacorporation

As we can see, the management of the knowledge required to ultimately deliver services to consumers is greatly skewed by the organizational structure. In no cases, except perhaps for network development and management and finance, are the business units and the knowledge areas under the control of any one group of people. Instead, the knowledge that must be shared by these individuals is constantly being consolidated, detailed, shared, and distributed based on a complicated set of legal, organizational, and technological relationships.

In order to put together an overall knowledge management value chain, therefore, it will be necessary to create a separate value chain for each business unit (or line of business). In these cases, the analyst needs to piece together a trail that starts from the customer and follows backward through this labyrinth of organizational structure until a complete chain is built.

4.5.1.1 Developing the value chain for a geographically based business unit

Let's carry through another example of alignment analysis, this time using one of the geographically based business units of the large telecommunications firm. The company, let's call it *Stateside Bell*, has been in existence since the earliest days of telephone service, has approximately two million customers, and concentrates on providing consumers and small businesses with local telephone service.

In order to build a value chain for this company, we will start where one should always start with this kind of analysis, with the customer. The first question we ask is, "Given our ultimate telecommunications value chain, how can we trace backwards from the customer all the way to the beginning of the telecommunications value chain?" The first thing we need to do is to figure out which links of the chain are important, trivial, or not applicable. The second is to figure out how, organizationally, those values are being delivered.

It is important that we not forget at this point that Stateside Bell is part of a much larger telecommunications firm with many overlapping business functions that we will have to consider.

4.5.1.2 Tracing the value chain back from the customer

In the case of Stateside Bell, at least some of the analysis of the customer interface process is easy. Billing and customer service are handled completely by Stateside Bell's own customer services department.

Unfortunately, the sales process is made more complicated by the fact that Stateside Bell has no sales people of its own (they rely almost exclusively on customer service to provide that function), but the corporate-defined consumer support, business support, and pay phone support divisions each employ sales, marketing, and product development staff who fulfill those roles for Stateside

Bell (along with fulfilling similar roles for all of Stateside Bell's sister state-based organizations).

This means that some pretty significant parts of Stateside Bell's value chain are managed, controlled, and delivered by a group of people not associated directly with Stateside Bell's budgetary or fiscal responsibilities.

As we move backward through the value chain, we will find that the situation actually gets worse.

While the customer services department handles all billing operations, Stateside Bell is forced to use a software package provided by the parent corporation as "the standard" billing package. This means that while Stateside Bell controls the execution of billing operations, they do not control the way in which it is handled. That is done by a corporate group with a completely different set of priorities.

The network infrastructure maintenance for Stateside Bell is also handled by a centralized corporate group, though because of the geographic nature of this service, the group really does tend to act as if they were under the control of and a part of Stateside Bell itself.

Network development, the acquisition of the right to do business, and the creation of new products are all functions that Stateside Bell, in general, has very little need for and will, therefore, not show up in the Stateside Bell value chain, as shown in Figure 4.3.

Table 4.3 summarizes the organizational alignment issues for Stateside Bell as a consequence of its participation in the much larger telecommunications firm of which it is a part.

It should be very clear, based on this table, exactly how complicated the information and knowledge infrastructure of a telecommunications firm can get in these kinds of cases. What should also be obvious is that this kind of "schism-ing" of the organizational structure creates the potential for a great deal of knowledge dysfunction (places where the efficient transmission and processing of knowledge through the overall value chain gets perverted or fails completely because of the structure).

Figure 4.3 Stateside Bell value chain.

Table 4.3 Stateside Bell Organizational Alignment

Value Chain Area	Business Area Coverage
Billing	Stateside customer service
Marketing	Corporate consumer, business, and pay phone support divisions
Sales	Corporate consumer, business, and pay phone support divisions
Customer Service	Stateside customer service
Network Development and Maintenance	Corporate network operations and network support staffs
Creation, Acquisition	Not important functions for Stateside
Provisions, Activation	Corporate service order processing group

4.5.2 Aligning the value chain with the information systems

Obviously, if people, organizational structures, and knowledge were the only components involved in the business equation in this day and age, there would be no way that any business could survive. Too much information needs to be managed and coordinated by too many people in too short of a time for any company to function efficiently. No, there has to be some kind of glue, some kind of structure, some kind of knowledge and information management system that keeps everyone coordinated despite the organization's best efforts to keep that from happening. This magic glue that holds the knowledge value chain together is, of course, the company's computer information system.

Computer systems do a lot more for the business than many people give them credit for. Computer systems not only keep track of the innumerable amount of trivial detail that drives a technologically and detail-oriented business like telecommunications, it also schedules, coordinates, times, officiates, resolves conflicts, sets priorities, and issues reports that disclose how well or how poorly everyone is performing. There is an incredible amount of information and knowledge brokering that a typical corporate information system provides, and if we are going to comprehend how the corporate value chain really works today, and if we are going to figure out how to architect a future environment that will maximize those processes, then we need to understand how the information systems align with both the organizational structure and the value chains.

4.5.3 Kingpin systems: the beginning of computer systems alignment

Unfortunately for us, not only do organizational structures and value chains fail to align in a lot of ways, but neither do the computer information systems that drive these businesses line up with either of them. The computer systems infrastructure represents a third, even more complicated layer in the overall knowledge structure of the company.

Fortunately, there are several things about the way the telecommunications industry works that can help us to greatly simplify the process of figuring out how things should line up.

The telecommunications industry, and almost every other industry, has certain key operational characteristics about it that will dictate a lot about how the information infrastructure is going to be put together. While there will certainly be variations among various sizes, types, and ages of companies, in general, as you go from one firm to the next you will see certain patterns in how the computer systems are deployed.

For the telecommunications industry, there are several reasons for this.

First, the legacy of most telecommunications firms derives from the fact that most of the older companies in the business started as government agencies or government-sanctioned monopolies. That meant that telecommunications companies, in general, were not competitors and in many cases were part of the same, much larger company. As a consequence, information systems and databases tended to be propagated pretty freely between the firms. (How many U.S. telecommunications companies use the old ATT CRIS customer information system or a variation thereof? Almost all of them.)

Second, even if the firm is a new startup company, the people that staff it will, to a large extent, be ex-employees of the larger, older firms. Therefore, the "industry knowledge" about the best way, in fact the only way, to set up systems is propagated throughout the industry at a subconscious level.

Third, these ways of doing things obviously make functional sense for a lot of reasons.

In the specific case of the telecommunications industry, what we find, almost without fail, is that two particular operational management systems become the kingpins and the driving information systems centers of any telecommunications firms. These are the switching and billing systems.

Ultimately, almost any information of true value to the company is found in one of these two systems. And, of equal import is the fact that the interface between these two systems is the most important set of transactions to the survival of the company as a business unit (see Figure 4.4).

When we stop to think about it, this actually makes a lot of sense. In fact, we can perceive the entire telecommunications value chain as really having two major parts to it, the first being everything having anything to do with the creation and maintenance of the "network" itself (the "network side") and the second being everything having to do with the marketing, packaging, selling, and collecting of revenues for providing those services (the "business side").

We include in the network side creation, acquisition, activation, provisioning, network development, network maintenance, service order processing, and switching management. We include on the business side billing, sales, marketing, customer service, and other areas dedicated to working with and servicing customer needs.

It should come as no surprise then that within any telecommunications company we will find not one, but two almost autonomous information systems infrastructures. On the network side we find the provisioning, activation, service order processing, and other systems, all revolving around the switching infrastructure and all focused on maintenance of the delivery side of the business.

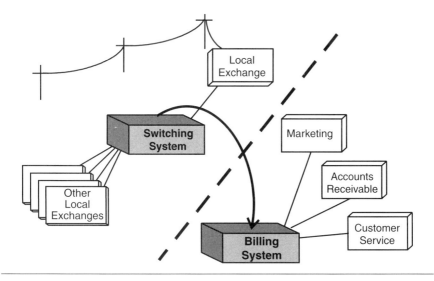

Figure 4.4 The "kingpin" systems behind telecommunications firms (billing and switching).

The real "heart" of the environment consists of hard-line connections to the consumer's home, which then converge together at a series of secondary switching points, which eventually consolidate to the "exchange" level and ultimately to the centralized network switching offices (in the traditional consumer phone service industry). In the cellular industry, the network infrastructure will consist of "repeater cells" that are all connected to central collection and routing points, which are ultimately tied into a centralized switching activity collection and consolidation point.

In general, any other systems on this side of the business get their input from some point along this chain.

On the business side, the starting point of all activities is the billing system. The consolidation of transaction detail on the network side of the house is formatted and transported to the billing system, which then uses the information to create customer invoices and activity records.

Almost all other systems on the business side of the house use information from the billing systems as the starting point.

4.5.4 Alignment problems and their symptoms

The incredible reliance that a telecommunications firm has on the billing system as the beginning and the end of all business information can obviously create problems for that business from a knowledge management perspective. In fact, a large percentage of the knowledge management problems that telecommunications firms suffer from today can be, to a great extent, blamed on this very problem.

To better understand how this occurs, we will need to take a look at the two things that happen to operational systems over time that turn them from nice, clean, functional, and efficient deliverers of operational value into the large, complex, complicated, incredibly expensive, and almost unmanageable messes that most telecommunications information systems are today.

4.5.4.1 Hybridized operational systems

When a telecommunications firm starts up, there will be a relatively clean and stable computer system environment consisting of a switching and a billing system, each of which have been designed do their respective jobs. Over time, however, and sometimes even immediately, people realize that a few modifications to the core systems code can make some business or switch management function more useful to them.

On the switching side of the house, for example, network planners will quickly realize that they need to have some way to keep track of and monitor the traffic that is flowing through the billing system in order to make better network capacity planning decisions. That would require a way to "tap into" the efficient flow of information through the switching system in order to capture and manage that information. So, as a consequence of this insight, either a new system for monitoring and managing network flow is created and the switching system is modified to send data feeds to the new system, or the switching system itself is modified to perform this same function.

In the first case, the new system being added will be incredibly expensive because new computer hardware, software, infrastructure, and so forth must be added. In the second case, the tacking on of a few little extra functions seems to make a lot of sense and will cost very little.

Ultimately, what happens over time is that more and more functionality is added to the switch management side of the house. New systems are added and tied in or new functionality is built into the switching system itself. Either way, the end result is a conglomeration of systems with no clear operational or knowledge management function. A nightmare has been created.

On the business side, things are actually even worse than this. We start with a nice simple billing system. Its job is to capture, consolidate, and prepare statements of service utilization for each customer—a relatively straightforward task. Unfortunately, the same kind of short-term efficiency problem takes hold on this side as well.

For example, marketing decides to offer customers different kinds of call packs and different billing rates to be applied at different times of the day, or for calls to particular groups of people, or any other long series of conditions and qualifications. All of a sudden the billing system is not just doing simple tallies of call units anymore. It has also become an extension of the marketing department.

Not only that, but now the finance, accounts receivable, and customer service areas also need to become involved.

The billing system needs to keep track of customers. So, before you know it, the billing system becomes the customer management system.

As long as we have all of the billing and customer information in one place, why don't we get the billing system to keep track of the history of customer service complaints at the same time? That way we can do a better job of keeping track of customer satisfaction.

And the process goes on and on.

Before you know it, the billing system *becomes* the company. Nothing happens that the billing system isn't driving. Obviously, what starts out as a good idea becomes a nightmare as the system continues to grow and change.

Each generation of enhancements to the billing or switching systems makes those systems more difficult to change, more expensive to run, and most catastrophically of all, more critical to the functioning of the business.

4.5.4.2 Operational systems as knowledge stores

As if the hybridization and wholesale interdependency of operational systems functions was not enough of a problem, we run into a second set of problems as time goes on.

Not only is it convenient to try to turn relatively simple operational systems into incredibly complex ones, but it makes even more sense to try to make these operational systems the primary vehicle for the communication of knowledge between business units.

When we turn operational systems into knowledge storehouses, we really complicate matters. All of a sudden we start wanting the operational systems to keep track of history.

On the switching side, we will realize that keeping track of the history of call pattern traffic for the past three years can provide us with an incredible source of knowledge about what the characteristics of our customers and their habits are.

On the billing side, we discover that keeping track of all of a customer's old bills is a good way of helping figure out if they are good customers or not. So, we modify the system to keep track of that information as well.

Eventually, we end up with mess that is so complicated and confusing that we become totally paralyzed. We can no longer afford to manage any more information because there is no place left in which to slip a few more pieces of data.

4.5.5 Data warehousing as an alternative

The condition we have just described is one that most telecommunications firms find themselves in. They usually have information systems that are large, complex, and devoid of any one set purpose, focus, group of people to serve, or knowledge function to support. The systems actually take on identities all their own, identities that follow their own rules and establish their own criteria for their use and modification. And it is into this environment that we propose data warehousing as an alternative.

Specifically, one approach to take with our data warehousing efforts is to consider the warehouse environment as an alternative to this highly interdependent, almost nonfunctional information systems soup that we have created.

What we can do is take a look at any and all of those situations where the information processing that has to happen is *not part of the operational mission of the system*, remove that functionality from it, and place it into the warehouse. Now, obviously, this is easier said than done, and there is no way to simply convert everything into that environment overnight, but the eventual migration of all nonoperational information into a nonoperational knowledge repository makes a lot of sense:

- It will help us to resimplify the operational systems and to focus and tune them so that they can manage their operational tasks better.
- It buys us all kinds of "headroom" for the development and exploitation of new kinds of information and knowledge management and creation systems, which can help the company achieve even higher levels of efficiency than they are currently capable of.

4.5.6 Data warehousing as a migration path

However, as we said, there is no way to magically transform the existing information systems infrastructure into this kind of environment overnight. The critical operational, informational, and knowledge-sharing tasks that the current information systems smorgasbord provide are critical to the company's success. Moreover, the wholesale creation of brand new warehouse repositories without concern for their practicality and short-term profit enhancement capabilities is a guaranteed formula for failure. What we need is a plan for the gradual conversion from the operationally dependent to the more balanced operational/knowledge-based world, one step at a time.

4.5.7 The fully aligned model—a summary

The solution, then, is to develop a plan that envisions the alignment of the information systems, the organizational structure, and the customer value adding functions as a comprehensive environment. The process will not be easy. In fact, it will probably and ultimately prove to be physically and organizationally impossible, but what the previous discussion provides us with is some insight as to how we might get started and where we can envision it going.

The starting points are as follows:

1. Develop a meaningful value chain for each line of business to be included.
2. Develop an understanding of how the organizational structure aligns with and supports the ultimate process of meeting customer needs.

3. Develop an understanding of how the current information systems align with, help, and hurt the process of delivering that value, based on their relationship to the business areas and their missions.

4. Develop a value-chain-based model for the ultimate warehousing environment that includes data storage areas that can support

   - The optimization needs of each autonomous functional area of the business (knowledge-based warehouse tables specific to marketing, sales, customer service, network development, etc.);
   - The cross-business area optimization needs that involve the mixing of information from different parts of the value chain (a value-chain-based "horizontal" warehouse, whose purpose it is to provide a physical environment where the cross pollination of knowledge will occur).

5. Finally, develop a plan for the deployment of a physical environment to support them all, which includes an approach that allows each component of the system to be added on a cost-justified-only basis.

Wow, that may seem like a tall order, but in the following two chapters we will be addressing each of these issues in turn. In the next chapter we will talk about how to set up the physical infrastructure to meet these demands, but doing so one step at a time and on a cost-justified basis.

In Chapter 6 we will look at the whole issue of value propositions in the telecommunications industry and provide guidelines and starting points for the identification and prioritization of opportunities that look like good places to start the process.

Chapter 5

Building the warehouse— one step at a time

So far, all of our discussions about data warehousing in the telecommunications field have been emphasizing the business, organizational, and knowledge management aspects of the system. These are all important, and we have shown how a good understanding of these aspects will contribute to the success of the overall warehousing efforts. But ultimately, all of the speculation about the usefulness of a warehouse will be lost if we do not have a good plan of attack for actually physically constructing it. At this point, any of the classical approaches to developing your computer systems environment will immediately start talking about databases, networks, and other architectural characteristics. But, just as we need to be sensitive to the organizational and existing computer systems environment so as to best employ the warehouse for delivering good information, so too must we take these things into account in order to make good technical decisions.

5.1 Challenges to infrastructure design

When it comes right down to it, the developers of any information system initiative like a data warehouse are faced with some pretty big problems when thinking about undertaking a project of this nature. The natural inclination, especially for the well-organized data processing professional, is to think in terms of putting the whole system together right away. This approach offers many advantages.

The primary advantage to building the entire warehouse at once is that if you can figure out everything that it needs to do before you get started, you will be able to build an incredibly cost-effective solution. There are many reasons for this:

1. *Leveraging hardware and software costs*—It is always more cost effective to place as many operations as possible on the same hardware platform. That is precisely why mainframe computers are as popular as they are. It's a simple matter of physics and economics. When you centralize operations, you reduce your expenses not only on the "hard" technology side, but on the "soft" costs side as well. The soft costs include systems management and maintenance, physical plant costs, and the cost of software like operating systems and databases. Not only does this ability to scale up reduce overall costs, it also makes it possible to handle volumes of data that are not possible on smaller machines. The problem of cost-justifying a machine big enough to handle high volumes is, of course, the big challenge.

2. *Planning and implementation costs*— It is believed that if you can figure out everything that will have to be done over the next three years, you can make a more comprehensive plan to manage the process effectively. In reality, of course, these kinds of long-range, large-scale planning efforts hardly ever work. Things are too complicated, and technology and plans change too quickly.

3. *Economies of scale in development*—This is perhaps the worst and least credible of the beliefs about systems development. Some people think that you can accomplish "assembly line" levels of efficiency in the deployment of large systems. This mistaken belief has proven itself to be false again and again in the history of I/T development. Systems development is, by its nature, a custom manufacturing process. Economies of scale are minimal.

In fact, much of the "science" of information systems development tells us that this is exactly what we should do. Among the disciplines that promote this approach are the following:

- *Enterprise modeling*—The belief that you can capture all of the "logical models" that drive the business and put them into the "perfect" dictionary. Enterprise modeling, while theoretically possible, has in fact proven itself to be functionally impossible.
- *The systems development life cycle*—This is the approach, born in the 1960s, that attempts to dictate what the right way to build a system is. It includes feasibility, analysis, design, implementation, testing, and production migration. The systems development life cycle turned out to be far from the optimal model for the construction of a data warehouse.
- *The Santa Clause Syndrome*—Probably the biggest detriment to warehouse development can be found in the very nature of the relationship between I/T and business people. Over the decades, I/T and business have developed a strange kind of relationship that I like to call the "Santa Clause Syndrome." In these situations, I/T personnel have become the keepers of the I/T resources and business people have become the wishers, asking for the things that make their lives more convenient, more comfortable, or more profitable. But I/T meets those needs based on how easy it is to provide them. Pretty soon, with report modification backlogs and unmet needs in a wide range of areas, the whole process of evaluating the effectiveness of the I/T solution gets lost. We lose the focus on what kind of business value we are going to deliver, and focus instead on how "happy" end users are.

Ultimately, the approach these disciplines advocate is one what we call the "dump and run" model of warehouse building, as shown in Figure 5.1.

Many, many organizations have tried to approach data warehousing in exactly this way, with abysmal failure rates.

There are many reasons why this approach will not work, and in fact cannot possibly work given the nature of the systems we are talking about building. Most data warehousing books, experts, consulting firms, and vendors will tell you that these approaches *are* viable, and that is exactly why people are having so much trouble making large-scale systems like this work. The reasons for the failure of this "dump and run" approach are many, and we list just a few of them here:

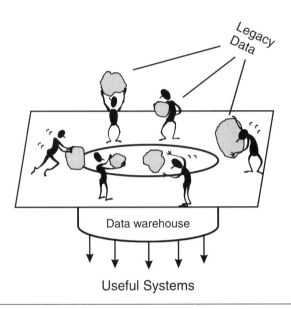

Figure 5.1 The "dump and run" model.

1. *They ignore the knowledge management paradigm completely*—When you try to approach warehousing using the logical modeling approach, where you build based on subject areas, organizational structure, or existing computer systems, you will end up with data structures that have minimal value in the short term and almost no value in the long term. Knowledge-based systems pinpoint the real areas of the business that warehousing can benefit the most effectively.

2. *They grossly underestimate the complexity involved in the identification and preparation of legacy systems data*—Historically, people who get into warehousing for the first time budget less that 20% of the time and money for the process of identifying and preparing legacy system data. They naively assume that all the data they need exists, and that it is easy to find and easy to use. Without fail, these people learn the reality, which is that over 70% of the budget needs to be allocated for this part. It's the ugliest, most difficult, and least glamorous of any job in I/T, but it is critical to the system's success.

3. *They fail to understand the shift in the relationship between I/T and business when these types of systems are deployed*—This is the failure of the OLTP paradigm (these are knowledge systems). Data warehousing

systems are not OLTP systems. They do not concentrate on the efficiency of existing processes. Data warehousing and data mining create new processes where there were none before. Users cannot give you requirements for a warehouse in the same way they can tell you how an order entry system should work. Data mining requires users to reinvent themselves and their jobs in order to be effective. Even more importantly, end users are not capable of giving you requirements for systems that involve the use of data mining tools that they have never seen or used before. Therefore, a new partnership-type relationship between I/T and business needs to exist—a partnership that involves I/T and business people working together to discover, develop, and deploy things that neither has ever done before.

4. *They fail to understand the "learning" nature of the warehouse environment*—Ironically, there are two ways data warehousing guarantees, if done correctly, that the developer will redo the job over and over again. First, the best warehouse implementation is one that creates new knowledge; that is, knowledge that tells the business person that something about the business should be changed. The insight *should* result in the business changing its procedures and legacy systems. Of course, now we have to change the warehouse to adjust to those changes, and the cycle continues. Secondly, and even worse, is the fact that after users gain a certain level of information, if that information proves useful, the very next thing they will want is *more* information.

For all of these reasons it is impossible to undertake the building of a corporate telecommunications data warehouse as a big, one-time project. No, an alternative project is required.

How do we go about putting it together in a way that allows us to leverage the advantages of large-scale database servers and large-bandwidth networks, when we are forced to implement the system one little piece at a time? How do we prevent the environment from turning into a collection of hundreds or even thousands of little databases (miniwarehouses) on hundreds of little PCs strewn across the corporate landscape, and the concomitant problems of administration and efficiency that this scenario envisions?

The answer is that we must somehow come up with a way to plan far enough ahead so that we can make the big purchases early on in the process while still focusing on the delivery of immediate value to the business. The approach we use to perform this miracle is called the business-led architecture approach, and it makes use of some new techniques for the specification, planning, and deployment of a warehouse architecture.

5.2 The functional components of a warehouse environment

Before we can begin to illustrate for you the way this revolutionary new approach works, we have to establish some basic understanding and some basic vocabulary for just what exactly it is we are trying to build.

A data warehousing environment is made up of three component pieces: the *acquisition area* (identifying data in legacy systems and making it available for the warehouse to use), the *storage area* (databases and/or files that hold the data in a form that makes it easy for end users to get at), and the *access area* (includes the network, personal computers, and data access tools that end users leverage to work with the available information), as shown in Figure 5.2.

The warehouse consists of these three functional parts, but is also held together and supported by two critical infrastructure components: the *physical infrastructure* (the hardware, network, and software that hold the system together) and the *operational infrastructure* (the people, roles, responsibilities, and procedures).

The typical information systems architect will want to immediately start basing his or her decisions about that infrastructure on the physical parameters. But that is actually impossible to do. In reality, you are never really going to be able to stabilize the physical infrastructure at all. What we submit to you here is that the real way to make an effective warehouse is to concentrate on developing

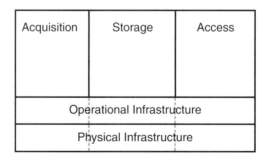

Figure 5.2 The warehouse and its component parts.

the operational infrastructure as the first priority, and simply "cobble together" the physical infrastructure as best you can at any given point in time.

The following sections will detail some of the critical information and characteristics about each of these components and will help illustrate exactly how we come to this uncomfortable, but necessary, conclusion.

5.2.1 Acquisition

Acquisition is the largest, most difficult and complicated, least glamorous, and most critical of all of the warehouse components. We include within our definition of acquisition all of the legacy systems of interest and all of the ways we can find the critical information they hold and make it available for use within the environment.

As much as we would like to believe that this is a trivial area, that is far from the truth. First of all, we must deal with the fact that most companies are forced to live with an incredibly diverse and incongruous set of legacy systems. Mainframes, minis, PCs of every size, brand, and description are usually cobbled together into some kind of unmanageable mess that we know as the legacy systems environment. In order to make the warehouse useful, we will have to figure out how to get the information out of each of these and place that extracted data into a database that is usable.

Any attempt to stabilize or create the one ultimate warehouse architecture will immediately be thwarted by these facts about the legacy systems environment. For one thing, the legacy systems environment is always changing. Therefore, the warehouse can never be more stable than the legacy systems that feed it.

As we stated earlier, acquisition represents 60% to 80% of the cost of warehousing, and 80% of that cost is the investment in people and their knowledge about the legacy systems themselves. Unfortunately, while many companies are trying to sell "automated" data extraction tools that are supposed to make this job easier, so far they have failed to deliver as promised.

The reality of data acquisition, at least for the present, is that it is a process that is mostly about the writing of programs that extract, format, cleanse, merge, purge, and otherwise prepare data for loading into the warehouse. This usually involves the creation of a long series of programs, each of which do a different part of the job, as shown in Figure 5.3.

The process of acquisition is actually made more complex by the fact that these operations are hardly ever performed on the same platform. The problem of building an acquisition job stream, therefore, includes figuring out which parts of the process should run on which machine (see Figure 5.4).

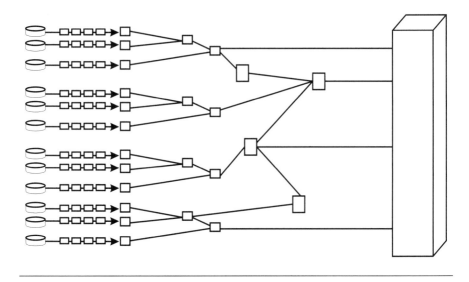

Figure 5.3 Acquisition (data preparation) job streams.

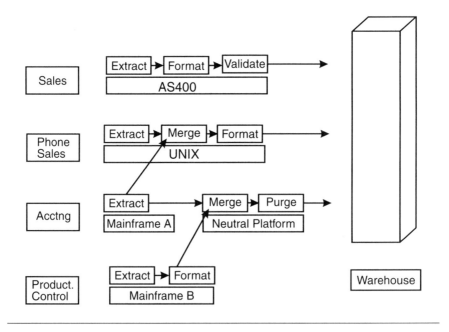

Figure 5.4 Acquisition on multiple platforms.

5.2.2 Storage

The storage area of the warehouse is probably the easiest to understand. To meet the storage needs of the warehouse environment, we really need nothing more than to provide a database or a file storage area, or even just a tape drive that can hold the formatted data, and make it available to end users.

Several things of special interest about the storage area include the following:

1. *It doesn't have to be a database*—Tapes, flat files, databases of all sizes and shapes can all effectively meet the warehouses storage needs. What's important about this area is that it store data in a form that is usable. As long as users can get at it, it's good enough.

2. *It doesn't have to be limited to one database*—Today's client-server, middleware, and distributed database technology make it possible for us to envision and create a megastorage area, made up of several databases, running on various kinds of hardware and using different kinds of database software. The assumption that to be an effective warehouse we must have *one* database can create a lot of unnecessary complexity.

3. *Data basements versus data warehouses*—One of the biggest mistakes a developer of a warehouse can make is to assume that the databases being built are going to be permanent and relatively stagnant. In the worst cases, companies have built warehouses to hold many years worth of historical information with no clear idea of who was going to use it. The assumption is that someday someone will want it. Then the information sits there and is never used. We call these kinds of systems, those that store data for the sake of storing it, "data basements." Data warehouses, on the other hand, need to be like real warehouses. You don't use a real-world warehouse to hold items that nobody is going to use. That's what museums are for. No, the only way to measure the effectiveness of your warehouse is to figure out how often and by how many people it will be used. A data warehouse should be measured, just like a real warehouse, in terms of how many "turns" the data makes within it. And just like the real warehouse, its manager should be ready to eliminate any inventory item that isn't getting enough use. Adaptation of this attitude means that the warehouse manager will need to be constantly vigilant and always ready to drop tables that aren't being used enough and add tables that can deliver new value.

5.2.3 Access

In many ways, the development of the access area of the warehouse is the easiest part. Data mining tools relieve I/T of most of the process of application programming and design. For the most part, these tools simply "hook up" to the warehouse and immediately let the user start being productive.

The technological piece of the process, the hooking up of the PC to the warehouse, is almost always managed using normal client-server technology and middleware. So, that, at least, is relatively straightforward.

What can be confusing, however, is the dizzying array of different tools and capabilities that are available to do the access job. While almost all products attempt to integrate different features and functions into their offerings, we have identified seven different major categories of tool type/functionality that all of the products fall under. When evaluating the particular tools, determining first which types of tools and which particular functionalities you are looking for can make the process a lot easier.

5.2.3.1 The seven access tool types

These function/tool types include the following:

1. *Query and reporting*—Traditional query managers and report writers;
2. *Agents*—Software that schedules, runs, evaluates, or searches for things for the user;
3. *OLAP (online analytical processing)*—Multidimensional analysis tools;
4. *Statistical analysis*—Traditional SPSS/SAS and other stats packages;
5. *Data discovery*—Neural networks, CART, CHAID, and other advanced artificial intelligence and knowledge-generating software;
6. *Visualization*—Systems that graphically or geographically display complex data relationships;
7. *Web tools*—Software that performs search, query, and agency work in the WWW environment.

5.2.3.2 The challenges of getting requirements

Given a good understanding of what kinds of tools are available is, of course, only half of the problem. The other half comes when trying to help users figure out what to do with the tools—to do that we need to go through an education/indoctrination process where we demonstrate the tools and help the users decide how they can best leverage them to solve the business problems they are facing.

5.2.4 The operational infrastructure

Far more important then any of the technical details required to make the warehouse work is the operational infrastructure that we provide for it. Hardware, software, legacy systems, and business needs are guaranteed to change, and to change often and violently. What will not change will be the need to have people who understand the legacy systems and how to get information out of them, who understand the databases (what's in them and how they work), the business users and the nature of their problems, and data solutions.

The operational infrastructure—that collection of people, skills, experience, policies, and procedures—is the way you tie the warehouse together and make it usable.

5.2.5 The physical infrastructure

It is in the area of the physical infrastructure that most technicians want to spend most of their time planning and executing. While these issues are important, they must take a back seat to the dynamic needs of the business and the technology. By definition, the acquisition area will be in a constant state of flux. The cost of storing data and the size of the optimum data storage area is shifting and dropping drastically on a regular basis. The power and ease of use of access tools guarantees that this environment will also be undergoing change.

The sad truth is that the optimum technical solution today is guaranteed to be less than optimum tomorrow. Therefore, the way we develop an optimum physical infrastructure is to remake that decision each time we add functionality, dynamically changing the environment to meet the needs of the business today and in the future.

5.3 The step-by-step, cost-justified approach

Given all of this undependability in the warehouse environment, the next question you might have is, "How do I go about building one?" The approach that we will be talking about here has been used effectively by many organizations in the past.

The approach assumes that the way to gain efficiency and to optimize your warehousing efforts is to envision the process as one of building the warehouse as a collection of application layers, as shown in Figure 5.5.

Using this plan, we create an initial physical and operational infrastructure to support one to two years worth of development. We then identify the best value propositions we can find and prioritize their development. Eventually, as

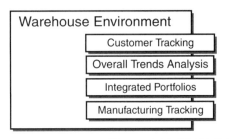

Figure 5.5 Application layers—leveraging efficiencies between applications.

the environment matures, it becomes possible to leverage more and more of what exists in the environment, driving the cost of each new application down lower than the one before it.

5.3.1 What is a value proposition?

Any data warehouse is, in fact, a collection of the value propositions it supports. But what exactly is a value proposition?

For our purposes, we refer to a value proposition as any contiguous collection of acquisition, storage, and access components which, when utilized together, deliver some kind of value to a group of end users. The whole topic of specifically what kinds of value propositions there are for telecommunications companies, and how to deliver them, will be the subject of our next chapter.

5.3.2 Gathering value propositions

Given that the first step of any warehouse construction project should be the identification of a collection of value propositions, and given the challenges most I/T departments have in getting their users to help identify them, we have found the following approach to be useful.

We view the technique of gathering value propositions as actually being a three-part process.

5.3.2.1 Step 1: vision development

The first step is to assemble groups of users from related business areas and hold what we call vision development sessions. A vision development session has three parts.

First, you spend some time talking with the users about the business problems they are currently trying to solve. Are they worried about customer turnover and churn? Are they trying to cut overhead costs out of the network infrastructure? Are they trying to control the cost of doing business? After discussing and developing a consensus among all of the attendees (from the business and I/T side) about what the most important problems are, you begin the second part.

During the second phase of vision development, you acquaint business users with the many different kinds of data access tools available and you show them how these might be used to solve the business problems. Depending on the technical and business sophistication of the users, and on the nature of the problems you are trying to solve, you will need to explore the use of various types of tools and at a varying degrees of depth.

For example, some users may be well acquainted with statistical analysis tools; so there is no reason to talk about those. Many will be completely new to OLAP products. Seeing how one works may encourage them to find some new ways of addressing their problems.

The final phase is to work with the users in developing their own specific value propositions. Make sure that they elaborate not only on what and how, but insist that they tell you exactly what kind of business value that particular functionality is going to deliver.

5.3.2.2 Step 2: organization and prioritization

After gathering a wide variety of value propositions of different business value, complexity, and business focus, we are ready to begin the next step of the process, which is to analyze them all.

During value proposition analysis we determine the following:

- What value each value proposition has;
- How difficult it will be to identify and obtain the data required;
- The extent of infrastructure investment needed to deliver the value proposition;
- How easy or difficult it will be to do;
- What leverage value it will have for other value propositions (If you do this one first, will others be able to make use of the work already done?).

After analyzing each value proposition, we will then be in the position of being able to prioritize and organize them into a logical deployment order.

Experience indicates that we should look at delivering warehouse "applications" in three-month blocks of time.

Given this prioritization, we can then estimate disk, database, network, and warehouse staffing budgets far into the future (one to two years) with some accuracy.

5.3.2.3 Step 3: development and construction

Finally, all we are left with is a list of value propositions that we can now begin delivering as soon as we have put the concomitant hardware, software, and operational infrastructure in place. See Figure 5.6 for a diagrammatic overview of the entire approach.

5.4 How do you build a warehouse?

So, for those of you who picked up this book looking for that checklist of "how to build a data warehouse in five easy lessons," here are the steps in summary:

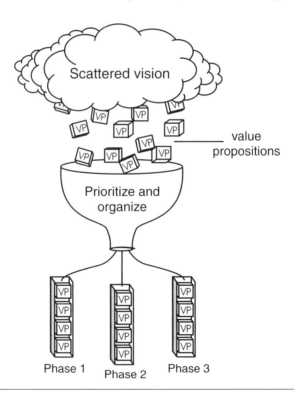

Figure 5.6 The value gathering approach.

1. Gather value propositions.
2. Prioritize and organize them based on the business value offered, the level of difficulty and cost to deliver, and the cost in time and money that the deployment of the new warehouse environment will engender.
3. Build the physical and operational infrastructure based on how best to deliver those value propositions as cost-effectively as possible.
4. Deliver the first value proposition.
5. Deliver the next value proposition.

While this business-led approach to warehouse architecture may seem less than efficient in a lot of ways, it does provide the developer of the warehouse with a roadmap that, when combined with knowledge management analysis, provides a firm, cost-justification-based plan of attack for some time to come.

Chapter 6

Value propositions in telecommunications

ULTIMATELY, the value of the data warehousing efforts that any telecommunications company puts together will boil down to the sum total of the values of the individual value propositions. The warehouse itself is nothing more than a framework and an environment within which the value propositions are run, and the data mining tools are nothing more than the means to put them on the user's desk.

But how does the combination of data mining tools, warehouse data, and business problems turn into meaningful value propositions, and just what are they?

One way to get a handle on this is to consider the various categories of data mining tools we mentioned in Chapter 2. Each one has a different type of value it can deliver.

6.1 Mining tools and value delivery

The seven types of mining tools are query, agents, OLAP, statistics, data discovery, visualization, and web based. We gave you a brief description of each in Chapter 2, but obviously, there is more to these tools than simply different ways to do similar things. Each has been developed to uniquely meet the information processing needs of business people.

No matter what the tool, however, and no matter what the type of data, value propositions fall under one of two major categories. We call the first category *operational monitoring and control* and the second one *knowledge discovery and exploration*.

6.1.1 Operational monitoring and control

Value propositions fall under this category when they are concerned with providing specific information about how the business is running and what is happening, where, when, how, and why. The key here is that the business person already knows what he or she wants to know, and knows more or less where to find that information.

A wide variety of value propositions fall under this category, including many that might seem rather boring. Operational value propositions provide users with answers to such questions as, "How many new customers did we put online yesterday?", "How many customers have been delinquent in their payments?", and "Is this customer a chronic complainer, or is this concern about service legitimate?"

Operational value propositions answer these kinds of questions using several formats:

- *Query and reporting tools*—Any time a user needs to know how well the business is running, he or she will first turn to a report or a query screen for the answer. The query and reporting tools represent probably more than 50% of all value propositions executed within data warehouses anywhere in the business environment.

- *Online analytical processing tools*—OLAP tools are another extremely popular instrument used to aide business people in analyzing the state of their business today. The OLAP tools work best when used to solve problems that involve very large, complicated, multidimensional data sets. These tools allow users to navigate down through the data, slice and dice the results, and look for disruptions in the patterns of the flow of business.

- *Agents*—Agents can help business people with monitoring by scheduling reports for them, scanning those reports, and letting them know when there are unanticipated problems.
- *Web tools*—Web tools function in much the same way as query tools do, the only difference being that they allow the user to ask questions from and about documents and presentations as opposed to databases.

6.1.1.1 Data sources for operational monitoring

In Chapter 3 we talked about the different kinds of warehouses that might be built. There were historical and real-time warehouses, and single-silo and cross-silo types. In general, the operational-monitoring-type value propositions will depend more on the real-time warehouses than the historical ones, but at times the historical type can provide some value as well.

6.1.2 Discovery and exploration

As we said, the major characteristic of an operational value proposition is that the user has some advance idea about what information he or she wants. With the knowledge discovery and exploration value propositions, the opposite is true. These value propositions, and these tools, are designed to teach and show the user things about their business and their customers that they did not know before. In the discovery and exploration area, we try to get the answer to questions like

- What is similar about all of the people who are good, high-profit customers?
- What is the estimated revenue potential for a new market we are considering acquiring?
- How can we tell our good potential customers from our bad ones before we invest in approaching them?

As with operational value propositions, these questions can be answered in different ways as well:

- *Statistical analysis tools*—These tools are used to describe and define major characteristics and to discover predictable patterns that can help the business make better decisions in the future. Using the traditionally and mathematically proven techniques of averaging, correlation analysis, regression analysis, factor analysis, and others, end users are able to combine mathematical intelligence with raw data to develop complex answers to very difficult questions.

- *Data discovery tools*—These tools attempt to answer all of the same kinds of questions as the statistical analysis tools, but use newer, more advanced, less mathematically proven techniques. The objectives for tools like CHAID, CART, neural networks, and the entire parade of artificial intelligence and nonstatistically-based mathematical proofing techniques are to answer the tough forecasting, anticipating, predicting, and explaining of human, machine, and system behavior without requiring the prerequisite mathematical savvy and experience. These tools make advanced statistical analysis as easy as plugging in the numbers and interpreting the results.

- *Visualization*—Visualization products allow users to see complex information displayed in ways they never saw it before. As problems become more complex and as the patterns being searched for become more obscure, one-dimensional questions and two-dimensional reports simply fail to provide us with the input we need to understand the situation clearly. Visualization tools help change that.

6.2 Value propositions by functional area

Every functional area of the telecommunications firm, every link in the company's value chain, can have its profitability, efficiency, and effectiveness enhanced by the application of data warehousing and data mining technology. Of course, while every area can benefit, we want to be certain that we start with those that are the most important, where the most value can be applied. This is precisely why we emphasized the development of a value chain and why it is important that we include in this value chain only those functions that are critical to the company at that point in time.

We will now consider only a few of the many hundreds of value propositions that telecommunications firms are discovering and cashing in on around the world. We will look at these propositions within each area, but will start with the more popular and immediately and obviously profitable ones, and then proceed to the less common ones. Each firm's priorities and needs are different, and this ranking does not indicate that some of the less common ones might not be the most critical for your firm. They just haven't shown up as often across the range of all firms.

6.2.1 Marketing value propositions (historical/cross-silo/discovery)

By far the most popular and most successful data warehouse/data mining combinations to have an impact on telecommunications have been in the marketing area—specifically, warehouses that hold historical information (our first category of warehouse) and that integrate that historical information from several sources within and outside of the firm (a cross-silo-type application). The analysis performed is of the knowledge discovery and exploration type (our second type of value proposition category).

Specifically, what telecommunications companies have discovered is the following:

1. They have a lot of information about their customers (who they are, where they live, when the use the phone, etc.).

2. They are facing an extremely competitive environment. Understanding who their best customers are and figuring out how to identify them, communicate with them, make them customers, and keep them happy is the major survival strategy of the next decade for many of these firms.

The key tool is what is known as a marketing database or a customer-based warehouse. We will spend a significant amount of time in later chapters talking about this phenomenon and seeing some of the different ways that data mining tools are helping these companies to accomplish these objectives.

In general, a combination of knowledge discovery analyses is most commonly used, statistical analysis and data discovery products being the most popular. With these tools, businesses identify their best customers and target their marketing activities to reach them.

Information from billing, customer service, call detail, and other systems are all pulled together and used to help marketing people develop their marketing profiles.

6.2.2 Credit value propositions

While not as glamorous as the marketing value propositions, credit management can get a big boost in productivity and effectiveness through the use of warehouses. There are actually two ways that warehouses have been applied effectively.

6.2.2.1 Credit value propositions (real-time/cross-silo/monitoring)

Customer monitoring systems, which keep track of how much service each customer is using and track that against their history of activity and their credit risk profile, make it possible for credit departments to anticipate delinquent or troubled accounts, reducing the debt exposure of the firm by a considerable amount. These systems use real-time data (daily or weekly updates) and query or reporting tools to constantly monitor the activities of customers and to flag trouble accounts immediately.

6.2.2.2 Credit value propositions (historical/cross-silo/monitoring)

Special historical analysis data warehouses, which hold demographic, historical activity, and other forms of customer profile information, actually allow credit analysts to create predictive profiles about what types of customers are liking to get into trouble and when.

One company noted that when phone service dropped off by more then 20% and then continued to drop off for an extended period of time, 35% of the time this was an indication that the customer was having financial problems and was soon going to be in a position to no longer be able to pay his or her bills. By identifying the pattern, the company was able to create pre-emptive programs that either provided temporary credit assistance to the customer or convinced them to take a lower level of service offering, thereby saving a bad debt situation and creating a loyal long-term customer.

Predictive models of credit risk cases have been built my many firms to develop profiles, which are then used by marketing (in the creation of special programs for each group) and sales (to indicate which customers to pursue and which to leave for the competition to have).

6.2.3 Customer service value propositions—(real-time and historical/cross-silo/operational monitoring)

While credit and marketing have had the most success with the historical/discovery combination, customer service is also finding great value in effective real-time warehouses that provide customer service people with information at their fingertips about all contacts the company has had with the customer, their billing history, their satisfaction rating, the ads they have received, and their current ranking as a customer (high profit, low profit, high risk, low risk, etc.). By giving customer service personnel this information, they are able to make better decisions about how to treat the customers and how to make the best use of the contact time that they have with them.

6.2.4 Sales value propositions

Sales units have found all sorts of interesting uses for mining and warehousing to help them in the execution of their tasks.

6.2.4.1 Sales performance tracking (historical/single-silo/monitoring)

Most sales organizations find it extremely difficult to keep track of how well each account rep is doing. While certain base measurements of salesperson activities are captured by the billing and service order processing systems, it is often troublesome to reconstruct the activity base for all the reps and to know how good a job they are doing. For example, how many of salesperson A's clients remain good customers for more than a year? How many of B's customers are collections problems? And so forth. Special sales tracking warehouses, when combined with OLAP or query tools, make this kind of analysis easy.

6.2.4.2 Channel management warehouses (historical/cross-silo/discovery)

In fact, in those cases where the company has multiple channels to manage (direct sales, retail outlets, agents, etc.), a historical, cross-silo warehouse with information about how each of the channels is working and how their sales relate to the profitability of the customers can save the company millions of dollars on bad channel decisions.

6.2.5 Network planning value propositions

The network planning area may find itself greatly in need of useful information to plan future network infrastructure. There are several ways that warehouse data can help.

6.2.5.1 Network infrastructure design (real-time or historical/silo/visualization)

One of the biggest challenges for network planners is developing a good understanding of what the competitive horizon and the current traffic patterns (activity levels, loads on different sections) look like.

Visualization tools make it easy for engineers to see exactly what is happening in ways that no printed report can communicate. In some cases, products are used to watch current traffic patterns and flows through the lines. Different colors represent distinct loads on each section. In other cases, historical call detail data is fed into geographical displays to pinpoint what kind of call activity is occurring and where.

Cellular and PCS engineers use visualization tools to help isolate repeater and cell traffic problem nodes. Trouble spots and failed service reports are posted to a geographical mapping system, which includes the position, type, and orientation of each node. This combined data can be used to shore up flagging links in the network.

6.2.5.2 Engineering and statistical activity analysis (historical/single-silo/discovery)

Many times, engineers will be able to tap into call or even switch activity detail and use advanced statistical analysis, or even operations management type approaches, to assist them in making network configuration decisions.

6.2.5.3 Traffic engineering—new network design and the use of historical information (historical/cross-silo/discovery)

Traffic engineering is by far the oldest (and still an incredibly useful part) of the telecommunications firm's arsenal of tools in the competitive battle. When asked to develop a configuration for an all new network area, engineers are in desperate need of input data. Any information that can help them anticipate what the traffic load might be will assist greatly in designing a more efficient network. In these cases, historical warehouses holding call detail information, cross-referenced with marketing and customer activity data for a geographical area similar to the one being targeted, can be used to help engineers extrapolate about what will need to be done to service the new area.

The science of traffic engineering was pioneered by A. K. Erlang, a Danish engineer (often referred to as the father of teletraffic theory). Erlang first applied advanced statistical and predictive models to the problems of defining telecommunications traffic problems.

The basic unit of measure in a traffic study is known as a CCS or 100 (C) calls (C) per second (S). You measure the traffic that is flowing through the network in terms of CCSs or Erlangs. (An Erlang is 36 CCS. Since there are 3,600 seconds in an hour (60 × 60), 36 CCSs = 1 hour of usage.) The traffic engineer tries to figure out the network capacity required based on the number of CCSs or Erlangs the network must carry at any point in time. In fact, most regulatory tariffs include a specification for the service level a carrier must provide to their customers.

Service levels are measured as the probability of receiving a "network busy" signal when you pick up the phone at any time. The grade of service level is stated as P.xx, where xx is the percentage of probability of the busy condition. A P.05 indicates that when you pick up the phone, there is a 5% chance that you will not be able to make your call at that time. Most tariffs require a grade of service of P.01 or better. Obviously, the ability to compute, predict, and guarantee this level of service means that the traffic engineer must have access to good data about the traffic flowing through the system and access to statistical analysis tools that can help him or her to make accurate predictions.

6.2.6 Network maintenance value propositions

Network maintenance can benefit from many of the same types of systems that network design uses to help them anticipate trouble spots and demand on their resources. However, this particular area has been getting a lot of help in the optimization of their process, namely, the efficient deployment of resources to address problems and the ability to anticipate problems and perform pre-emptive maintenance activities.

6.2.6.1 Customer service call routing and scheduling systems (real-time/single-silo/visualization)

One way customer service groups have cashed in on warehousing has been to create visually based central clearing warehouses for all service-related activities. Using these facilities, dispatchers can get a "birds-eye" view of where customers are having problems, the nature of those problems, and where the closest available service personnel can be located.

This real-time analysis activity can actually provide the dispatcher with the ability to look for patterns in the occurrences and allows for some crude diagnostic capability in the process.

6.2.6.2 Service rating and ranking systems (historical/single-silo/discovery)

When all service call activities for all service personnel, all types of calls, and all response times are captured in the same place, it becomes possible to analyze the efficiency of different personnel and to look for newer and better dispatch and coverage strategies.

6.2.7 Creation

The telecommunications firm with a heavy need to invest in the process of creating new products and services and identifying the best markets to deliver them to can benefit the most by providing access to the marketing database types of systems that we have mentioned previously. Identifying the patterns of consumption, and the acceptance patterns of other markets for similar or related products, can help the service creator discover facts that can only enhance his or her ability to create the new types of products.

6.2.8 Activation and provisioning and service order processing

The activation and provisioning areas of the business are mostly concerned with the efficient execution of activities based on the demands put upon them by the rest of the business. These groups can often tap into marketing, customer service, and billing warehouse systems, and, using the same base data but different kinds of reporting tools, make the processing of efficiently executing and even accurately anticipating the demands of the system a lot better.

Systems that track the progress of service orders as they travel through the labyrinth of departments, activities, and approvals that many larger telecommunications firms create can help everyone do a better job of figuring out how well people are doing their jobs and how well the customers' needs are being met.

One telecommunications firm found that it had an incredibly large number of complaints about slow response times in requests for new lines. A service order process tracking system soon uncovered the fact that the organization had built several bureaucratic bottlenecks that guaranteed late response times. The system was then used to help encourage everyone to "keep the customer first."

6.2.9 Billing (historical/single-silo/discovery and monitoring)

We left billing, and the warehousing capabilities to support, for last. This is because in most telecommunications firms, the billing system is considered to be, and often functions as, the main warehouse in support of billing as well as a lot of other functions.

In some cases, however, this burden becomes too much for the billing system to handle by itself. In these cases, a separate billing information warehouse is established. This system holds copies of recent as well as historical billing activity. Combined, this information can provide a lot of valuable input to the other business areas (we have already mentioned several cases where historical billing information could enhance marketing and other functions) and is a rich source of analytical input for the billing department itself.

6.2.10 Operations

The name "operations" is given to different functional groups of people in different types and sizes of telecommunications firms. Therefore, we will not even bother to try to cover how they can leverage use of the warehouse here.

Clearly, any operations group will be able to make use of the different functional areas we have mentioned depending on their particular makeup.

6.3 Conclusions

The value propositions we have mentioned here represent only a small cross section of the many ways that creative telecommunications companies are cashing in on warehousing approaches. As we have seen, while warehousing is an appealing proposition, there are a lot of things to consider before attempting to deploy one.

6.3.1 Knowledge management approach

We have proposed that the would-be telecommunications warehousers hold off on the construction of the system until they figure out certain key things:

1. A solid understanding of the existing information systems, their functions, their hardware/software makeup, their effectiveness, and their infrastructure is required. A warehouse can only function as well as the systems that feed it, and feeding the warehouse is 80% of the cost.

2. A good grasp of the organizational structure, the business roles and responsibilities of each group, the information systems they rely on, and their current critical issues must be developed.

3. We propose the development of a good, high-level corporate value chain. This value chain will tell the would-be warehouse designer what the core business functions are that contribute to the ability of the firm to deliver value to the customer. It will provide the identification of "knowledge silos," places where systems, people, and knowledge will tend to cluster and isolate itself from others within the organization.

4. We firmly stated that the warehouse's value is measured as the sum total of the value propositions that it supports. It is imperative that anybody considering these kinds of activities should start by identifying as many high-value, good return on investment (and hopefully easy to execute) value propositions as possible. In this chapter we have reviewed many of the major types of value propositions that other telecommunications firms have already been able to capitalize on.

5. Finally, with all of this input information available, we should then, and only then, proceed with the development of plans that will lay out the

required hardware, software, data mining tools, and deployment schedules for an ongoing warehousing effort.

In the following chapters we will be looking at data mining tools and value propositions and see exactly how these are executed successfully, using real examples from real companies.

Chapter 7

Simple sales analysis: an introduction to operational monitoring using Microsoft Query

Assuming that the telecommunications organization in question has been able to get all of the pieces together to create a viable warehousing environment, we are ultimately left with a fundamental question. *Now that we have a warehouse, what do we do with do with it?* We will be dedicating the rest of our book to answering this critical question. The way you decide to answer this question will determine the ultimate success or failure of your overall warehousing efforts.

From the broadest perspective, the objectives for deploying a warehousing solution should be tied to the overall strategic direction that the firm has decided to take. You may recall from Chapter 1 that we established three major and fundamentally different strategic directions a telecommunications firm can choose to take in order to attain competitive advantage. A company can choose to pursue (1) customer intimacy (making it the mission of every employee to figure out ways to get to know customers better and turn the firm into the best meeter of consumer needs in their market space), (2) operational efficiency (figuring out how to make your firm the most cost-effective provider of services in the industry), or (3) technological excellence (establishing a reputation for being the industry's leading-edge firm).

It should come as no surprise that most consumer-based telecommunications firms (traditional, cellular, PCS, etc.) tend to emphasize the first direction, while the "back room" providers (long lines, switching, etc.) tend to favor the second, and the telecommunications manufacturers lean in the technological excellence direction. In all three cases, warehousing and mining can provide a big competitive advantage if deployed correctly.

We also established in the early chapters that there are really only two kinds of business objectives you can use the warehouse to address. The first is to use the warehouse to help the business increase its efficiency in the monitoring and control of operations (these objectives are of value to all three types of strategic objective, but have the most import for the firm concerned with operational efficiency). The second was to provide for the ability to do advanced analysis and discovery (also useful to all three groups).

Finally, we created a basic foundation for understanding how these approaches can be made the most useful to the firm by developing a good understanding of the different functions performed as a part of its particular value chain. Without question, these tools and approaches can be applied with equal legitimacy to any functional area of the business.

Because there are so many different variations that we could talk about, and because we do not have enough room to provide the reader with detailed examples of all of the different ways data mining and warehousing have been and could be used, we are forced to take a representative sample type of approach. In the following chapters we will explore several of the foundational concepts and applications of these approaches as they apply to specific telecommunications company cases, while at the same time highlighting the general principles that these solutions represent, thereby making it possible for the reader to see how these types of solutions could be applied to other areas.

Because the telecommunications industry is so highly specialized, we will make use, in all situations, of real-world case study models that should make the examples immediately recognizable to the reader. Many of the cases have been somewhat modified and disguised to protect the anonymity and competitive advantage that the different types of solutions imply.

We will start, then, with the simplest type of case, providing the reader with some basic understanding about how these solutions are put together from the business, organizational, informational, and technological perspectives.

7.1 Operational efficiency—an overview

The first area we will consider is that of operational efficiency. When we refer to operational efficiency, we are talking about all of the various ways that we can assemble information from within different operational systems and make it available to business decision makers in a form that enhances their ability to understand how well the business is running.

The main objective of any operational efficiency value proposition is to help people monitor how well the parts of the business are running, identify existing problems, and anticipate future ones. This helps us to understand what should or should not be a part of the scope of an operational monitoring activity. In general, of course, knowing how well things are running can be a fairly useless exercise if the business person looking at it can do nothing about it. Therefore, the biggest criteria for what we deploy as a part of an operational efficiency value proposition is that the information reported be *actionable*. In other words, you only report things to people who can do something about it when they discover problems. Actionable information can then be used to help the company prevent or minimize losses and increase revenues in a variety of different ways.

Looking at this basic premise in terms of the value chain of the telecommunications firm, we can see some obvious applications in the following areas:

- *Provisioning and activation*—Making sure that the process is proceeding as scheduled;
- *Sales*—Measuring the effectiveness of sales people and channels, and adjusting those channels as needed;
- *Customer service*—Keeping customers happy by being able to answer their questions about the status of their service and problems;
- *Finance and accounting*—Keeping track of all of the financial metrics that drive the business;
- *Billing*—Tracking and reporting on the accuracy of bills, and making it possible to research historical problems and their resolution.

Since we cannot provide the reader with examples from all of these areas, we will begin by concentrating on some specific cases relating to the monitoring and control of the sales and channels management processes.

7.2 Sales monitoring and control

This is an area that all telecommunications firms have in common and an area that is also relatively simple and easy to understand. The problems revolving around sales tracking, as they relate to its real-world application, might seem at first puzzling to the uninitiated outsider. How hard could sales monitoring be? You've got salespeople, you've got some kind of sales budgeting and forecasting system that tells you what salespeople should be selling, and you've got the billing system that tells you what has been sold. Sales tracking should be a simple matter of pulling together information from these sources and making it available to everyone.

Unfortunately, things are never that simple. While there are forecasting, budgeting, provisioning, and billing systems in place, each of them was developed to produce different kinds of reports. These reports met the needs of the sales tracker when they were first developed, but over time things change. Quotas are modified, compensation plans are varied, and the way the business runs changes dramatically. These changes in the business, however, are not usually reflected in the operational systems that run it. While businesses tend to change sales policies and procedures frequently, information systems can only be changed very slowly and at great expense. The end result is that before too long, nobody can really tell anymore what is doing what.

A data warehouse can provide a solution to this problem. Instead of basing our need for sales information on the operational systems themselves, what we can do instead is make copies of the data that are pertinent to the reports that are needed, and make the current view of the sales environment available through a collection of data tables. The warehouse provides the perfect solution to this perplexing problem. As the business continues to change sales policies at a breakneck pace, the warehouse can continue to keep up by simply changing the types of data that it pulls into the sales tracking area.

Of course, just because the correct sales tracking information has been stored in the warehouse, that does not automatically mean that our problems are solved. If we tried to create production-type reports, or to customize online reporting systems in order to pass this information on to the management and sales staffs, we would very quickly re-create the same problem. To make the warehouse as responsive as possible to the many changes we know will be occurring, what we need to do is include the use of a query or reporting tool that the end users can utilize on their own. That way, as the business conditions change, the salespeople and managers can immediately change their reports, providing them with instantaneous, accurate, and reliable information.

7.3 A universal problem

The example we have just cited, which talks about the problems of keeping operational tracking systems in synch with the structural and procedural changes in the business, is hardly limited to the sales area alone. It is a recurring and endemic problem that literally every area of the telecommunications business is forced to struggle with on a regular basis. Businesses, especially telecommunications businesses, are part of an extremely dynamic environment. They must change quickly and effectively in order to survive, and data warehousing, when combined with simple, effective user-based reporting tools, can make the difference between the success and failure of that business in the long run.

7.4 Using Microsoft Query and Excel to do sales tracking

To provide the reader with a basic understanding of how this approach works, we will start with a very simple example, using the most basic of query products. We will show how this kind of sales reporting can be accomplished using Microsoft's MS Query.

While far from being the most sophisticated query product on the market, MS Query does provide the user with a good set of basic functionality. Furthermore, it is integrated into the Microsoft Office toolset, which means that users can employ MS Query to find the data they want, and then have that data automatically populated to spreadsheets, documents, and other databases. For example, to gain access to your warehouse and pull data from an external database into the spreadsheet using Excel, all you need to do is select the Data...Get External Data option from the menu bars (see Figure 7.1).

The MS Query product will automatically be invoked; and after you are through making your selection, the data will be automatically pulled back into the spreadsheet.

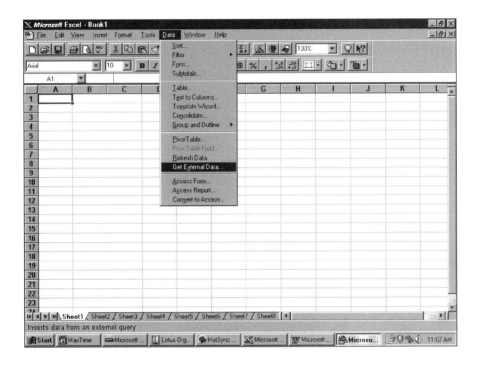

Figure 7.1 Invoking MS Query from within a spreadsheet. (Screen shot reprinted with permission from Microsoft Corporation.)

7.4.1 The sales database

For this example, we will assume that the data warehouse environment has been organized so that the needed sales information from the billing, accounting, forecasting, and budgeting systems have been organized and loaded into three tables, called *Sales* (holding all records of sales transactions on a daily basis), *Geographic* (showing sales reps and their respective branches, countries, and regions) and *Offerings* (providing information about all products and product lines), as shown in Figure 7.2.

Provided with this warehousing environment, the user can simply invoke the MS Query product to gain access to the data. The first thing you will see, once our MS Query environment has been invoked, is a query building screen, as shown in Figure 7.3.

This screen is made up of two major areas. On the top half of the screen we see all of the tables that have been selected for use in the execution of this particular query. On the bottom half are all of the specific columns and all of

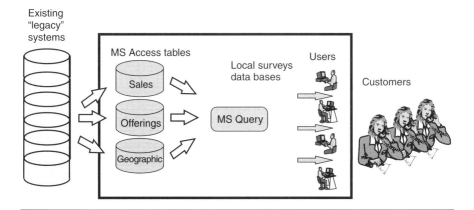

Figure 7.2 Data acquisition for sales example.

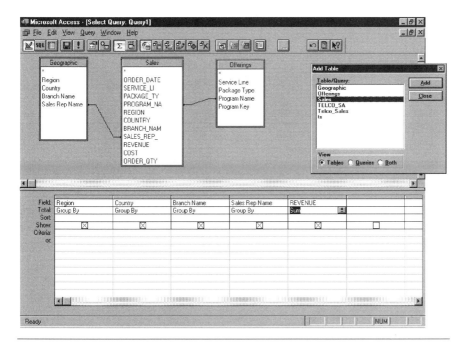

Figure 7.3 Selecting the tables for a query. (Screen shot reprinted with permission from Microsoft Corporation.)

the selection criteria (sort, merge, sum, and "select only if" conditions) that make up the actual query itself.

In Figure 7.3, you can see the following:

- *The table selection box (right side of the screen)*—This box shows all of the tables from which we might extract information.
- *The Geographic, Sales, and Offerings tables*—These are the three tables that we have chosen to execute this particular query. They were selected by simply double-clicking on the names of the tables.
- *The sales report criteria*—It shows that we are building a query that will provide us with a report that tells us the revenue each marketing rep has generated, broken down by branch, country, and region. These reporting columns were selected by double-clicking on the column names as they appeared in the selected table display boxes. (The Group By instruction, found in the Total: row tells the system to summarize everything up to that particular level, and the Sum instruction tells the system to add up all the numbers and report the total of all those values.)

Our successfully completed report can be seen in Figure 7.4.

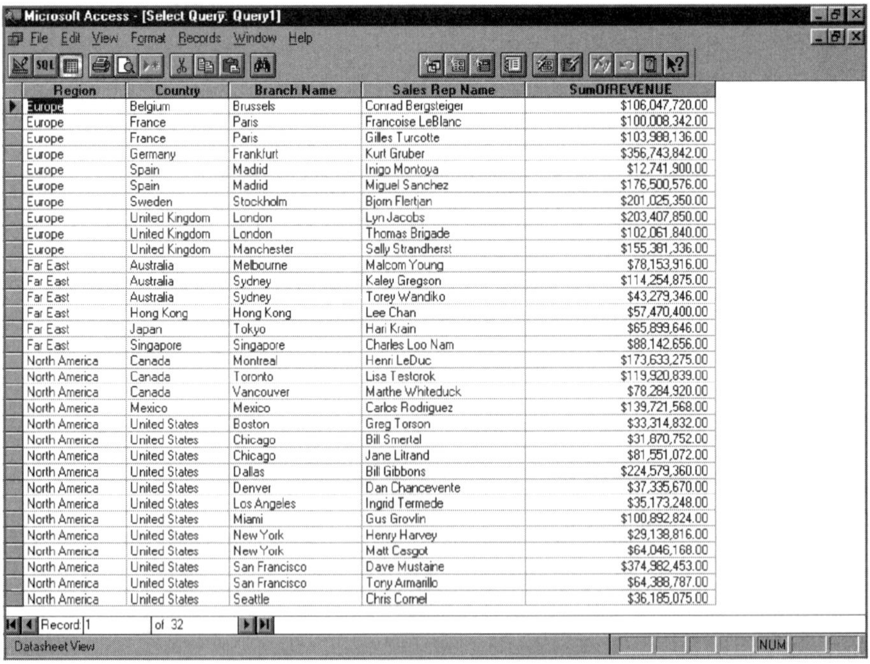

Region	Country	Branch Name	Sales Rep Name	SumOfREVENUE
Europe	Belgium	Brussels	Conrad Bergsteiger	$106,047,720.00
Europe	France	Paris	Francoise LeBlanc	$100,008,342.00
Europe	France	Paris	Gilles Turcotte	$103,988,136.00
Europe	Germany	Frankfurt	Kurt Gruber	$356,743,842.00
Europe	Spain	Madrid	Inigo Montoya	$12,741,900.00
Europe	Spain	Madrid	Miguel Sanchez	$176,500,576.00
Europe	Sweden	Stockholm	Bjorn Flertjan	$201,025,350.00
Europe	United Kingdom	London	Lyn Jacobs	$203,407,850.00
Europe	United Kingdom	London	Thomas Brigade	$102,061,840.00
Europe	United Kingdom	Manchester	Sally Strandherst	$155,381,336.00
Far East	Australia	Melbourne	Malcom Young	$78,153,916.00
Far East	Australia	Sydney	Kaley Gregson	$114,254,875.00
Far East	Australia	Sydney	Torey Wandiko	$43,279,346.00
Far East	Hong Kong	Hong Kong	Lee Chan	$57,470,400.00
Far East	Japan	Tokyo	Hari Krain	$65,899,646.00
Far East	Singapore	Singapore	Charles Loo Nam	$88,142,656.00
North America	Canada	Montreal	Henri LeDuc	$173,633,275.00
North America	Canada	Toronto	Lisa Testorok	$119,920,839.00
North America	Canada	Vancouver	Marthe Whiteduck	$78,284,920.00
North America	Mexico	Mexico	Carlos Rodriguez	$139,721,568.00
North America	United States	Boston	Greg Torson	$33,314,832.00
North America	United States	Chicago	Bill Smertel	$31,870,752.00
North America	United States	Chicago	Jane Litrand	$81,551,072.00
North America	United States	Dallas	Bill Gibbons	$224,579,360.00
North America	United States	Denver	Dan Chancevente	$37,335,670.00
North America	United States	Los Angeles	Ingrid Termede	$35,173,248.00
North America	United States	Miami	Gus Grovlin	$100,892,824.00
North America	United States	New York	Henry Harvey	$29,138,816.00
North America	United States	New York	Matt Casgot	$64,046,168.00
North America	United States	San Francisco	Dave Mustaine	$374,982,453.00
North America	United States	San Francisco	Tony Armarillo	$64,388,787.00
North America	United States	Seattle	Chris Cornel	$36,185,075.00

Figure 7.4 Revenue by sales rep report. (Screen shot reprinted with permission from Microsoft Corporation.)

7.5 Managing more complicated needs

Of course, the report example we have given here is a relatively simple one, and getting a report like this to work is not too challenging. However, as the business problems get more complex, we very quickly begin to identify some fundamental flaws in the straightforward query tool approach.

Some of the reasons that the simple query tools fail to be an optimal solution in many situations include the following:

1. *Complexity of the data*—Query tools are great when a user only needs to go after simple collections of data; but in most cases, truly useful information can be developed only when a large number of tables (sometimes dozens) are polled, resulting in large, complex query structures that a typical user is unable to develop.
2. *Complexity of the reports*—Not only is the accessing of data made difficult, but users usually require more complicated reports than what a simple query tool can provide.
3. *Complexity of the technology*—As the depth and breadth of reports increase, so does the amount of technological sophistication the users must have to get the information they need.

The simple fact is that the more complicated the reports get, the less useful the simple query tools are. In Figure 7.5, we can see an example of a moderately complicated report.

Just how useful is a report that is this hard to read?

7.6 Alternative methods of accessing data

Of course, simple query tools represent only the most rudimentary of techniques for the utilization of a warehouse environment. In future chapters, we will get a chance to look at some of the more sophisticated tools, and we will see how these can be used to address immediate business problems.

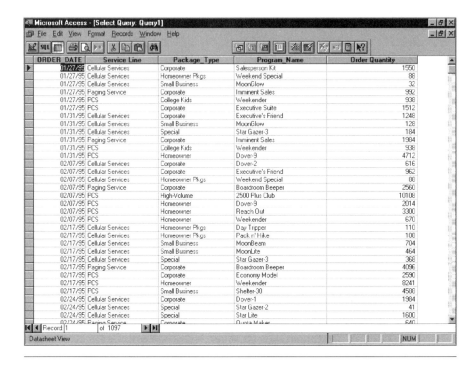

Figure 7.5 Sales report by date/service line etc. (Screen shot reprinted with permission from Microsoft Corporation.)

Chapter 8

Sales and product management: advanced operational monitoring using COGNOS PowerPlay

IN THE PREVIOUS chapter we introduced the reader to the fundamental concepts of operational monitoring and showed how a simple query tool can be used to help business people keep track of what was happening in different parts of the business.

Some of the basic principles we identified as key to the effectiveness of these solutions were as follows:

1. That whatever was being reported on be actionable (that someone could and would do something about the information provided);
2. That the information be easy for the business user to gain access to;

3. That the reporting format be easy to modify so that, as business conditions shift, the business person can vary the reports in order to keep up with the change.

Unfortunately, we found that, as the amount of information and the complexity of the reporting increased, the simple query tools quickly began to fall short in all of these areas. To address these more complicated kinds of situations, a whole new breed of sophisticated, powerful, advanced query and OLAP products have begun to flood the market. In this chapter, we will consider one of these, the PowerPlay product, provided by COGNOS. Using PowerPlay, we will see how users can be further insulated from data complexities while still gaining valuable insight from complex data configurations.

8.1 Monitoring complex business organizations

One of the biggest challenges faced by any business person today, is that of keeping track of the many different facets, organizational structures, and relationship dependencies that basically describe the business world of the late 20th and early 21st century. Businesses are huge and multifaceted, and the performance monitoring of the people, products, and services across diverse boundaries is a daunting task.

To help us to understand how this complexity can make things difficult, let's turn back to our sales tracking example for a moment. In Chapter 7's example, we found that it would be useful to develop reports that let us know how well different sales reps were performing in the offering of different kinds of telephony. We also found that we needed the ability to keep track of these sales reps across a wide range of geographical areas. When monitoring sales activity, we might want to know how well sales are doing in different regions, countries, or cities. We might also want to know how well our different lines of business are performing at several different levels. Obviously, expecting business users to construct all of these many different reports is unreasonable and would most likely discourage them from using the system at all.

Not only will each user want to look at information in a lot of different ways, but there are likely to be many users wanting to see it in ways specific to them. For example, sales information at a detail level is most meaningful to the sales representative. The sales manager for a given country, however, will be most concerned with those same sales numbers consolidated up to the national level. Upper management, on the other hand, will want to see those same numbers summarized to the regional level. The makers of PowerPlay and other advanced query-type products have come up with an alternative that addresses this kind of problem.

8.1.1 Determining the different levels at which to report

One way to make complicated matter easier for people to access and understand is to provide a simplified view of the matter. To accomplish this, someone needs to analyze all the information and predetermine the various ways users might want to see it. In the case of the advanced query tools, a data analyst investigates all of the many different ways people might want to view information and then figures out the different reporting hierarchies or "data dimensions" that people might want to use. Data dimensions define the lower, middle and upper consolidation levels at which a report may be viewed.

For our case study in this chapter, we will create a fictitious company called ABC Telecommunications International. ABC Telecommunications is a multinational firm with sales territories organized by cities, regions, and countries, with different sales reps reporting to different city offices. Since all of the sales numbers for all the reps in a given city can be rolled up into city, region, and country, we will identify this as our first reporting hierarchy and give it the name *geographic* for the Geographic dimension.

Of course, information about how well sales activity is proceeding is almost useless if we do not know what time period each of the sales pertains to. For that reason, we usually create a date/time dimension to allow us to track sales during various time periods. We will call this the ORDER_DATE dimension since we will be tracking all sales by the date on which the order was placed. Figure 8.1 shows an example of how the PowerPlay product displays our ORDER_DATE dimension, with data organized by year, quarter, and month.

We can now track sales by *how much*, *where,* and *when*. The information we still need is *what* was being sold. For this we will need to create a product *Offerings* dimension. In the case of ABC Telecommunications International, salespeople are responsible for the sale of a wide variety of different cellular, pager, and PCS phone services to corporations, small businesses, and directly to consumers. The different offerings usually involve the combination of special rates for service combined with promotional items, discounts on phone devices, and other incentives. ABC Telecommunications actually offers more that 40 different programs, all of which the sales reps can provide to their customers. Our Offerings dimension will therefore include breakdowns by service line (cellular, PCS, paging), package type (corporate, homeowner, college student, high volume, special, and other market segments) and different specific programs with names like (Star-Gazer, MoonGlow, Executive's Friend, etc.). Figure 8.2 shows a small portion of the overall Offerings dimension as represented within the PowerPlay product.

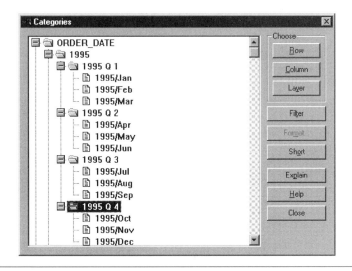

Figure 8.1 The ORDER_DATE dimension. (Reprinted with permission from Cognos Corporation.)

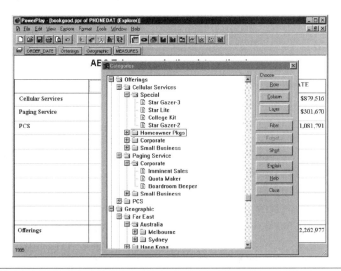

Figure 8.2 The Offerings dimension. (Reprinted with permission from Cognos Corporation.)

Once all of the dimensions for the report have been determined, the system designer is ready to make the new, easy to access reporting system functional.

8.1.2 Preparing the data for use

As we said earlier, the way to make complex data easy to get at is to simplify the process. While organizing the data into comprehensive dimensions is one part of that process, the other is to gather up all of the data and preconfigure it in a way that makes it easy to access. PowerPlay accomplishes this by creating what it calls PowerCubes. All of the data a user will need is gathered ahead of time and preloaded into the appropriate PowerCube environment (as opposed to pointing a simple query tool towards the correct database and leaving it for the user to figure out where the data is). This way, through the use of the PowerPlay interface, the users can navigate through the data in any number of different ways without ever having to know about the tables that lie beneath, as shown in Figure 8.3.

8.2 Exploring sales and product performance

After all of the data has been analyzed, acquired, prepared, and loaded into the PowerCube, we are ready to look at the new, friendly view of these complex relationships.

When users first enter the PowerPlay environment, they are immediately presented with a basic, high-level report, as shown in Figure 8.4.

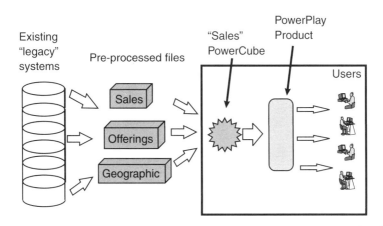

Figure 8.3 Creating the ABC Sales PowerCube.

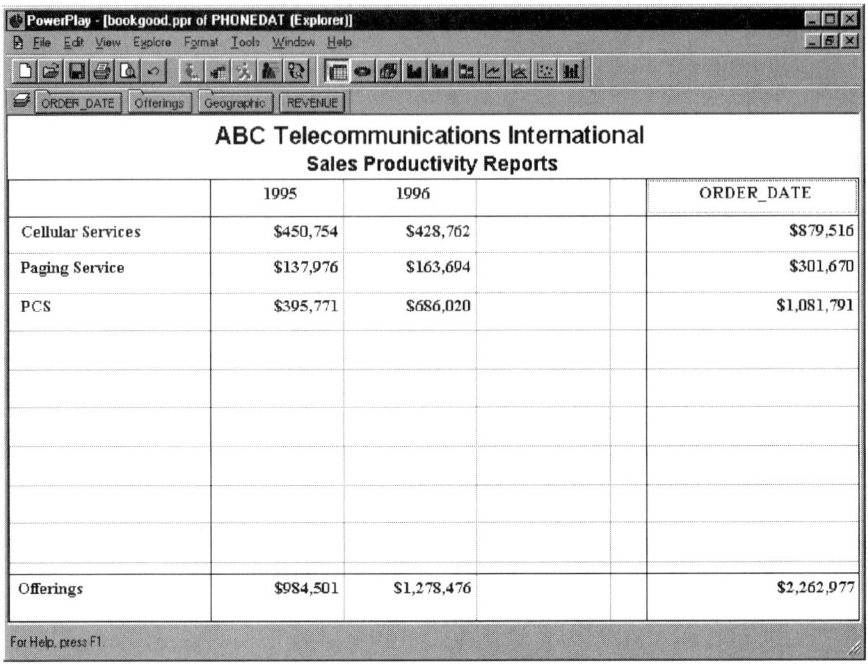

ABC Telecommunications International
Sales Productivity Reports

	1995	1996		ORDER_DATE
Cellular Services	$450,754	$428,762		$879,516
Paging Service	$137,976	$163,694		$301,670
PCS	$395,771	$686,020		$1,081,791
Offerings	$984,501	$1,278,476		$2,262,977

Figure 8.4 The base Sales Productivity Report. (Reprinted with permission from Cognos Corporation.)

You will notice that there are several small folder icons along the top edge of the reporting screen and you should recognize the labels on most of these folders. They are Offerings, Geographic, ORDER_DATE, and one that we haven't mentioned yet, REVENUE. The first three folders provide us with the means to search through different rollups of the reporting data, based on the Offerings, Geographic, and ORDER_DATE dimensions. The REVENUE folder is part of the *Measures* dimension, which allows us to decide on the numerical values to have summarized across the reporting dimensions.

You will also notice that the report actually shows us the revenue numbers for 1995 and 1996 for each of the three service lines and also summarizes the values in each row and column of the display. (See the far right and lower edge of the screen.)

Of course, this simple report in and of itself is nothing to get ecstatic about. What is exciting is what we can do from here.

So let's assume that someone in upper management has called up our Sales Productivity Report and is looking at these highest level numbers. Our executive notices that sales for 1996 are lower than for 1995 in the cellular services area. This concerns our executive, and he or she definitely wants to know the reason for the decline in sales. To investigate further, all the executive needs to do is double-click on the cellular services row. The row will then become highlighted, as shown in Figure 8.5.

The next thing on the screen is a report of sales from within the cellular services line of business only, broken down by the four different package types it contains (Special, Homeowner Pkgs, Corporate, and Small Business), as shown in Figure 8.6.

Our executive now immediately notices that although the Special and Homeowner Pkgs lines have shown increased sales from 1995 to 1996, falling sales are due to low numbers in the corporate and small business areas. At this point, the executive could choose to continue the analysis further by clicking on the Corporate row to see how the product sales for each program show up, or he or she could choose to navigate a different dimension. In our example case, the executive decides to see a little more detail about how those sales of cellular products pan out across the four quarters of 1996, the problem year. A

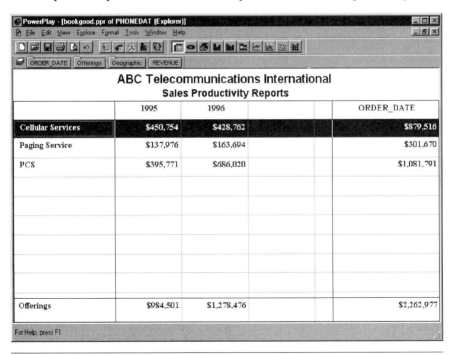

Figure 8.5 Clicking on Cellular Services for drilldown. (Reprinted with permission from Cognos Corporation.)

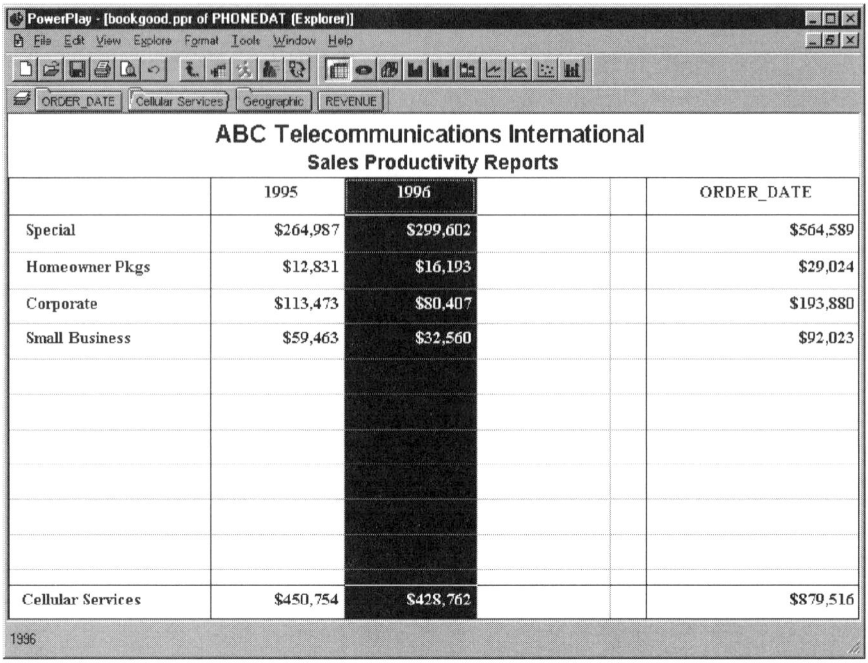
PowerPlay - [bookgood.ppr of PHONEDAT (Explorer)]
File Edit View Explore Format Tools Window Help
ORDER_DATE
Cellular Services
Geographic
REVENUE
ABC Telecommunications International
Sales Productivity Reports
1995
1996
ORDER_DATE
Special $264,987 $299,602 $564,589
Homeowner Pkgs $12,831 $16,193 $29,024
Corporate $113,473 $80,407 $193,880
Small Business $59,463 $32,560 $92,023
Cellular Services $450,754 $428,762 $879,516
1996

Figure 8.6 Cellular sales by program type. (Reprinted with permission from Cognos Corporation.)

simple click on the year 1996 (see the highlighted section of the screen in Figure 8.6) results in producing the same report, but this time broken down by the four quarters of 1996, as shown in Figure 8.7.

At this point, it should be quite obvious exactly how powerful and easy to use the PowerPlay interface is. Our executive can simply click on different reporting characteristics from any of the dimensions or measures available in the selection folders across the top of the screen and investigate all sorts of different perspectives on sales performance.

8.3 Additional PowerPlay features

In addition to the ability to navigate up and down across these dimensions, PowerPlay allows end users to instantaneously view graphical representations of the reports as well, allowing them to get a better "feel" for exactly what they are looking at. This is accomplished by clicking on one of the graphical selection icons located across the top of the screen, at the right. (See the bar with figures

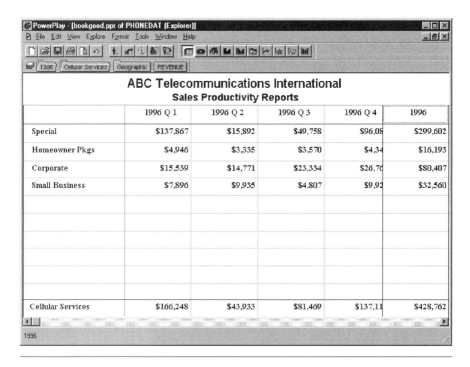

Figure 8.7 Breakdown by quarter. (Reprinted with permission from Cognos Corporation.)

representing pie, bar, plot and other types of display charts.) Figure 8.8 shows a pie-chart display of the Sales Productivity Report broken down by quarter.

What is nice about this graphical display feature is that the user can navigate up and down the dimensions viewing different graphical displays of the different levels by merely clicking on the piece of the graphical display that they want to see broken down further.

8.3.1 Alerts

PowerPlay, like most advanced query products, includes several other features that making using it even easier and more efficient. One such feature is the *alert* capability. Alerts allow the user to instruct the system to check for certain conditions within the body of any reports it produces, and when those conditions are noted, to immediately perform some other actions.

For example, our sales manager might set up an alert to check for any time that sales from this year begin to drop below sales for last year during any given week. When that drop in sales is noted, the manager could instruct the system

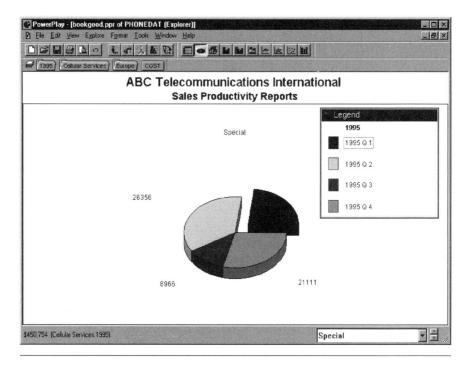

Figure 8.8 Pie chart by quarter. (Reprinted with permission from Cognos Corporation.)

to immediately send an e-mail message to the sales person responsible for the drop, letting him know that a problem has been identified, while at the same time sending the sales manager herself a report showing that rep's activity for the last four weeks.

Alert mechanisms are perhaps the most powerful enhancements to reporting systems to come along in some time. These capabilities can make it possible for a single business user to monitor the activities of hundreds of people, machines, or situations without ever having to worry about whether anything was missed.

8.3.2 Schedulers

Coupled with the alert capabilities of PowerPlay is the ability to let users actually schedule reports to be run whenever they want. *Schedulers* enable users to ask for reports to be run and audited for alerts on daily, hourly, weekly, monthly, or quarterly time frames, making it possible for people to get as many reports as they want, whenever they want, without requiring them to fill out forms, wait in lines, or do anything more than simply click on a screen.

8.4 Summary

The COGNOS PowerPlay product is one of several software packages that are currently flooding the market. These products are fast, efficient, easy to use, and they empower end users with capabilities never before imagined. Before you can make use of these OLAP capabilities, however, you must determine the dimensions and prepare the data and structure for the users.

Chapter 9

Customer intimacy: an introduction using SPSS

As we have noted many times before, most consumer-based telecommunications firms have adopted a major strategic initiative toward making their firms more intimate with their customers. While this may sound like a fine objective, it begs an important question: What does customer intimacy mean, and how can a company use warehousing and mining to achieve it?

In general, the size of the task faced by most telecommunications firms in trying to understand their customers is directly related to the firm's size and breadth of services. A typical telecommunications company has hundreds of products and services. How does a huge corporation like this understand who purchases their products and services? And how do they change their business practices in order to outperform their competition in terms of anticipating needs and becoming the preferred provider?

These questions can be daunting, but there are many ways that firms may approach these problems. Usually, solutions involve the following:

1. Increasing the level of customer service;
2. Creating more diverse and "customized" offerings packages;
3. Identifying what good customers want and figuring out how to supply their needs.

While there are many tactical solutions that can bring these initiatives to fruition, there remains the initial problem of understanding their customers, especially their likes and dislikes. Until these preferences are determined, any tactical approach to gaining intimacy will fail, for obvious reasons.

Fortunately, telecommunications firms have an enormous stock of data available about their customers. Almost no industry has as much data collected about the habits, preferences, and buying patterns of its customers than the telecommunications industry. Telecommunications firms typically know about every phone call each customer makes, including such information as when the call was made, how long they talked, to whom they talked, and how often they call back. They also know customer payment history, and in many cases can acquire assortments of related demographic and sociographic details through third-party market research firms or through the use of surveys, customer service polls, and response logs.

Clearly, if a telecommunications firm wants to understand their customers and their customers' needs, these firms have enough raw data to learn plenty. What is unclear is how exactly they can use the data. This is where warehousing and analytical data mining enter.

9.1 An introduction to analytical mining

While using query, agent, and OLAP-type tools can help the individual attempting to monitor and control business efficiency (as demonstrated in the previous two chapters), these tools provide little or no assistance in the more complex problem of understanding the behavior, habits, and predicted actions of people. To address these issues, we need a different approach and a different kind of tool set.

We have stated that in order to solve analytical problems, there are three types of mining tools—statistical analysis, data discovery, and visualization—that provide the requisite capabilities for a business person. In this chapter, we will look at some ways statistical analysis can be used to answer some of the many pressing questions telecommunications firms have about their customers' behavior. In the next chapter, we will see how a particular type of data discovery tool, the neural network, can be used to answer some other kind of questions along these same lines.

While we may concentrate on showing how these tools can be used to address marketing and customer intimacy questions, this by no means indicates that these are the only kinds of problems a business user can use these tools to address. For example, these tools can help find hidden patterns in company performance that might indicate an opportunity to reduce costs or increase efficiency. Or, they may help predict future behavior of customers, markets, employees, systems, departments, and so forth. Any time the business user needs to discover how certain things happen, these tools can be invaluable.

9.2 Statistical analysis—options and objectives

Statistical analysis is a branch of mathematics that derives knowledge from observed data. Statistics has a very rigorous and systematic basis in probability theory, but it is this probabilistic basis that makes the use of statistics as much an art as a science. Probability involves the concept of randomness in events, adding a degree of uncertainty to any event predictions. For this reason, statistics requires human interpretation, such as deciding whether a 95% probability of an event occurring is "good." It is this human interpretation that has made some people dubious about the use of statistics. However, if used appropriately, statistics can be a very powerful tool for business decision making, both in increasing productivity and efficiency as well as in growing revenues.

Statistics can be divided into two branches, "descriptive statistics," with which most people are familiar, and "inferential statistics," which requires a bit more training. Descriptive statistics are numerical or graphical methods of describing a set of data. For example, the average amount spent on telephone services by a telecommunications firm's customers is a descriptive statistic. Graphical methods such as plotting the frequency of calls made during different times of day are also descriptive statistics. Using descriptive statistics is often a good way to start a statistical analysis, since it allows you to look at your data in various ways before attempting to address specific questions about relationships.

Descriptive statistics are a good start, and they can be invaluable for presenting results, but there is an even more powerful tool for obtaining results from posed questions—inferential statistics. Inferential statistics uses principles of probability theory to find connections and relationships in data. For example, you might use your data to see whether there is a relationship between customer satisfaction with services and the amount of services purchased by individuals. There is unlikely to be an *exact* relationship between a satisfaction scale (e.g., 1–7) and the dollars purchased, but statistical methods can take a look at the variance in amounts purchased in relation to the variance in satisfaction levels to determine any trend. Methods such as regression analysis or analysis of variance (ANOVA) systematically parse out the variance in amounts purchased that can be attributed to customer satisfaction and provide a measurement of likelihood that some relation between the two exists.

Inferential methods often involve the practice of hypothesis testing, or analyzing data in an attempt to refute one possible theory in order to posit its antithetical theory. For example, you might refute the theory that there is no relationship between customer satisfaction and purchasing behavior and replace it with the knowledge that there is some relationship. This "knowledge," however, is a probabilistic knowledge. For example, you can say with 95% confidence that there is a relationship between customer satisfaction and amount of services purchased. It is up to the analyst and his or her audience to decide whether this 95% confidence is sufficient. However, knowing the level of confidence in your results is extremely important, and it is why inferential statistics is a good methodology when seeking knowledge prone to some level of uncertainty.

There is debate as to whether techniques such as neural networks can be construed as statistical methods. Neural networks are also based on mathematical principles, and in fact many of the algorithms used by neural networks were derived directly from statistical theory. However, unlike classical inferential statistics, neural networks do not use hypothesis testing. In fact, neural networks do not require a specific question about relationships other than, "Are there any relationships in my data?" For this reason, it is difficult to explain the results of neural networks, especially how individual variables affect other variables. This explains why neural networks are used most often in forecasting and prediction, when only a predicted outcome is required, and not a full explanation of the model. With regression, on the other hand, a user can analyze the influence of specific variables on other variables. In fact, regression has a systematic way for testing whether these influences are "significant." In other words, regression allows a user to make specific inferences about his or her data whereas neural networks simply look for patterns in data and, based on those patterns, predict future outcomes. Neural networks will be further explored in the next chapter.

9.3 Descriptive approaches

As we noted above, descriptive statistics are good tools with which to start any statistical analysis. Often, an analyst, especially one doing data mining, does not know much about the data other than which variables are included. In order to use company data to make general statements about customers and markets, it is helpful to first get a summary picture of the data before testing for any specific relationships.

Most people are familiar with the use of averages, whether from sports, school grades, or weather reports. Averages, which are sometimes called means or expected values, are the best guess at predicting the value of a variable with no other information. For example, suppose you want to predict the amount a new customer will spend on telecommunications services. Your best guess with no other information is the average level of expenditures by your other customers. So, if you are trying to measure the total dollar potential in a market and you know that your current customers each spend $100 on average, and you know that there are 500 potential customers overall (including your own customers), your best guess at market potential would be to multiply the average spent by your current customers ($100) times the total number of potential customers (500) to get a market potential of $50,000. Note that this assumes you have *no other information* about potential customers other than the number. For example, it may be the case that your customers spend more on telecommunications services than the potential customers do. There are other statistical tools that can help you use additional information about potential customers, such as demographic data from the census, to refine your measure of market potential.

A good way to refine the market potential analysis is to find different customer segments and use averages within customer segments to project opportunity dollars on those same segments among potential customers. For example, consumer database marketers often use a standard of living measure called socioeconomic status to segment current and potential customers. We could select appropriate socioeconomic segments, compute average levels of expenditures by customers within each segment, and project this average on the socioeconomic mix of potential customers. In this manner, if the socioeconomic mix of current customers is different from the socioeconomic mix of potential customers, using averages within segments is a better method than using the overall average. Projections of market potential are more accurate because you use a more appropriate weighting system.

There is a wealth of information on socioeconomic factors from U.S. census data. Unfortunately, often this socioeconomic data is split up into its component variables, such as income, education level, and job type. However, we can use a data reduction technique known as "cluster analysis" to use the information in three variables to create one summary variable. Cluster analysis is a descriptive statistical method that can find patterns in the values of certain variables for different cases in order to group together either variables or cases. Since we want to find socioeconomic groups for customers, we will group our cases (each case has information for one customer) using the variables income, education level, and job category. In SPSS, we simply select Hierarchical Cluster from the menu, choose the variables we want to use as information for grouping—including income, education level, and job category (see Figure 9.1)—and make sure

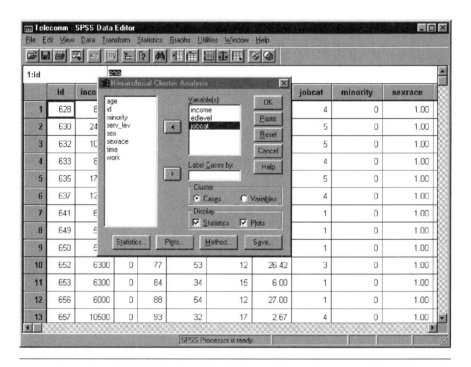

Figure 9.1 Running a cluster analysis in SPSS. (Reprinted with permission from SPSS Inc.)

we are grouping cases and not variables. We have also asked SPSS to create what is called a dendogram, which is shown in Figure 9.2.

The dendogram lists case numbers down the left-hand side and shows how each case groups with other cases over certain threshold measures of "distance," which is a composite measure based on the three variables. To the right, you can see that all the cases are eventually grouped together as we continue to relax this threshold measure. The dendogram gives a user flexibility in choosing the number of clusters based on the specific needs of the analysis. For our purposes, we will choose the three clusters circled. Of course, for most data, a dendogram like this is not feasible. However, you may use a dendogram on a subset to

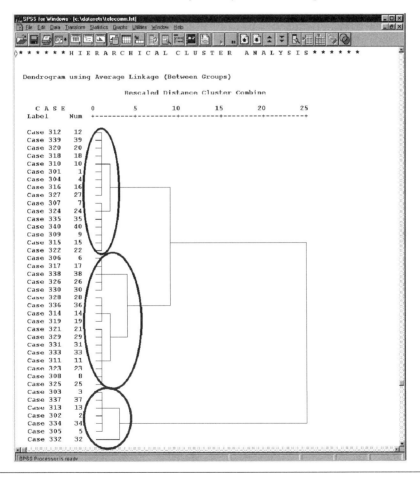

Figure 9.2 Cluster analysis results: dendogram in SPSS. (Reprinted with permission from SPSS Inc.)

decide how many groups you want and at what threshold level of "distance" in order to classify the entire data using numerical output instead. The groups can be named based on the predominant characteristics, such as "white collar," "blue collar," and "pink collar" workers. We can then look at the average expenditures within these groups for our current customers projected on the mix among potential customers to get a more refined measure of market potential. The advantage of this segmentation method is that the socioeconomic mix of potential customers may be different than the socioeconomic mix of our current customers, so we might overestimate or underestimate market potential by just using overall averages.

Although the average is our best guess at the likely spending of any potential customer, we might want to look at the entire distribution of spending by our current customers to see whether there is more information there. For this we can use a histogram, which chooses ranges of expenditures on the x-axis and plots on the y-axis the number of customers spending within this range. One such histogram is displayed in Figure 9.3. SPSS provides numerical statistics

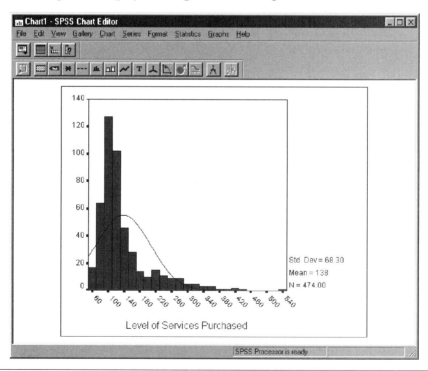

Figure 9.3 Distribution for level of services purchased: histogram in SPSS. (Reprinted with permission from SPSS Inc.)

such as the mean standard deviation (a measure of variation) and the number of cases to the right of the histogram. We can see that most customers are concentrated in the lower levels of expenditures. The distribution is not symmetric, meaning there are not as many customers spending a high amount as a low amount relative to the average expenditures. For this reason, we might want to get a sense as to what makes the high-expenditure individuals unique. For example, we might feel that the level of expenditures is related to certain characteristics of our customers.

There is a very powerful technique called regression analysis that can help us find how a dependent variable such as the level of services purchased changes with different characteristics of customers, such as age, income level, education, and gender. We can then target potential customers who have the "best" characteristics—those which increase the chance of the customer purchasing a higher level of services.

9.4 Inferential approaches—regression analysis

Regression analysis uses the variance in data to fit a line giving a predicted level of services purchased for any given set of characteristics. For example, we could predict the level of expenditures for a male, aged 29, who makes $30,000 per year. Regression analysis also provides probabilistic measures of fit and influence so that a user can know exactly how much "confidence" the user can have in the results. For this reason, regression is a well-established and widely used statistical technique.

To run our regression, we choose Linear Regression from the SPSS main menu and open the dialog box. We select the level of services purchased as the dependent variable, and age, education level, minority classification, sex, and income as independent variables, as shown in Figure 9.4.

The SPSS output is shown in Figure 9.5. The output gives an overall test of the model in the ANOVA section.

The second column from the right, labeled "F", shows the F-statistic for the regression. This is a statistic used to test the hypothesis (commonly referred to as the null hypothesis) that none of the independent variables used in the regression have a significant effect on the dependent variable. The alternative hypothesis is that *at least one* independent variable does have a significant effect on the level of services purchased. The right-hand column labeled "Sig.", which is called a p-value, gives the probability of getting an F-statistic at least this high if in fact none of the independent variables have an effect on the

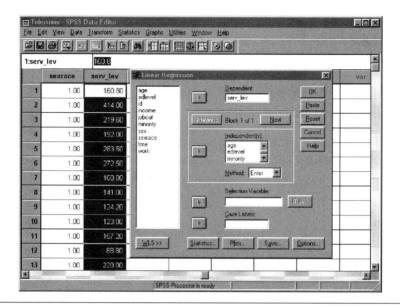

Figure 9.4 Running a linear regression in SPSS. (Reprinted with permission from SPSS Inc.)

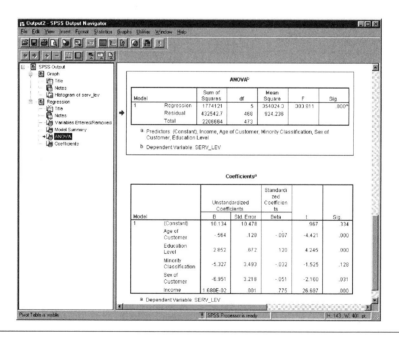

Figure 9.5 Regression results from SPSS. (Reprinted with permission from SPSS Inc.)

dependent variable. Our significance of .000 shows that it is very unlikely that all of the independent variables are not significant. We can therefore reject this null hypothesis and accept the alternative hypothesis that at least one of our independent variables has a significant effect on the level of services purchased. This means that we can use at least one of our variables to better predict the level of services purchased than we could by strictly using the average expenditures for all customers.

The section labeled "Coefficients" shows similar statistics for each independent variable separately. The column labeled "t" gives the t-statistic, and the column labeled "Sig.", which again shows p-values, shows the probability of getting t-statistics this far from zero if in fact there is no effect on level of services purchased due to this variable. We can see that for age, education, and income, the likelihood of no effect is very small. As a rule of thumb, statisticians typically use 0.05 as a threshold p-value. Any p-value below this is considered significant. Any p-value above is considered not significant. We can see that only the constant (which is the estimated level of purchases when all the independent variables are equal to zero) and minority classification of customers are not significant. The other independent variables are significant.

Besides having a measure of confidence in the effects from variables, regression analysis allows you to predict levels of services purchased from any values of the variables. The column in Figure 9.5 labeled "B" gives the coefficients for the line fit through the data. Our line is therefore (rounding to two significant digits)

$$\text{Level of Services Purchased} = 10 - 0.56 \text{ Age} + 2.9 \text{ Education} - 5.3 \text{ Minority Class} - 7.0 \text{ Sex} + 0.017 \text{ Income}$$

This means we can take any values of the variables and predict the level of services purchased by plugging these values into the equation. In our example, sex is coded 0 for males and 1 for females. Therefore, all else equal, a man is likely to spend $7.00 more on services than a female (the coefficient is negative, so a value of 1 subtracts the coefficient from the level of services purchased). Unfortunately, we cannot confidently extend a similar interpretation to the Minority Class variable since we found this variable is not significant in the regression. This means we cannot confidently say that whites are more likely to spend more than non-whites or vice versa. The coefficients also provide a good method for "What If" analysis. For example, we might ask, "If we have two people with the same characteristics except for the fact that one has $1,000 more income than the other, how much more is the higher income person likely to spend on services?" To answer this question, we simply take the coefficient on Income (0.017) and multiply it by $1,000, giving us $17.00. This means that a person with the same characteristics as another but with $1,000 more in income is likely to spend $17.00 more on services than the lower income individual.

Moreover, just as we can determine a level of confidence about certain variables having an effect on the level of services purchased, we can have a level of confidence about our predicted values. We have the highest level of confidence when we predict the average level of services purchased by our customers (which, from our histogram in Figure 9.3, we know is $138). This means that if we plugged in values for all the independent variables and got the number 138, we can be more confident in this specific prediction than in any other predicted outcome. As the predicted level moves away from this average, our confidence becomes less. In fact, there is an exact formula that allows you to compute the confidence level of your prediction for different predicted values.

Regression can also be used in forecasting. For example, some models look at sales next period as a function of sales this period. Regression finds trends through time and predicts future values of sales using past values. Not surprisingly, the confidence in our forecasts drops the farther into the future we look.

As you can see, regression analysis can be a powerful tool for finding relationships between variables. A user must keep in mind, however, that there are certain stringent assumptions being made when using linear regression, the most commonly overlooked assumption being the linear form. Relationships are often nonlinear and require more complex models. Still, a practitioner who understands regression methods well can be an invaluable resource by providing money-making solutions for businesses.

9.5 Conclusions on statistical analysis

It is ironic that the strength of statistical analysis lies in its uncertainty. Specifically, statistical analysis approaches uncertainty in a systematic manner, providing a level of confidence for statistical results. It is often hard to accept the existence of randomness in systems, people often calling this randomness or "error" a measure of our own ignorance about the way things work, but even in physics, a so-called "hard science," probability and randomness play a crucial role. The field of quantum physics developed out of this realization that randomness has a role in the universe. Therefore, by approaching randomness in a systematic way, statistics is not inconsistent with science, though we must remember that human interpretation plays a critical role. Other techniques, such as neural networks, do not require as much interpretation, but they also most often do not provide a measure of confidence in your results. We will discuss neural networks more in the next chapter.

Chapter 10

Predicting customer behavior: an introduction to neural networks

IN THE PREVIOUS chapter we saw several ways that traditional statistical analysis, when combined with the rich data resources of a data warehouse, can help telecommunications companies learn about the nature of their customers, thereby making it easier for them to figure out better ways to become intimate with them.

In this chapter, we want to take a look at an even more specific kind of challenge we face when trying to figure out how to work with customers and show you how a radically different kind of product, a neural network, can be used to do an even more powerful analysis of those customers.

While traditional statistical analysis may seem on the surface to be "magical" in many respects, its accuracy is well established and undisputed in academic and business circles. However, some problems get to be so big, or so complicated, that traditional statistical analysis is either unable to address them, or the enormity of the task presented is so large that the application of statistics to the solution of the problem is simply not economically feasible.

It is in these situations that we find uses for the newer generation of data mining tools, the data discovery tools. We include in this category systems based on techniques like CART, CHAID, decision trees, genetic algorithms, and a host of nontraditional mathematical analysis approaches. While the statistical analysis methods have hard mathematics as their basis, these products count on "softer" mathematics and "softer" proofs. Therefore, their use is more risky and speculative.

10.1 Unraveling complex situations

So, how can we go about predicting, for example, the future buying behavior of customers? How can we determine what the complete set of characteristics is that give us clues to their potential spending or turnover patterns? Statistical analysis techniques like regression, factor, and correlation analysis allow us to work with a limited number of variables, but the typical telecommunications firm has dozens or hundreds of different characteristics of a customer that might be taken into account. Are professional skiers better customers than restaurant owners? Are airplane pilots more likely to change service providers than school teachers? Are pet owners good or bad credit risks? All of these different characteristics might help us achieve better insights into who our customers are and what they need.

Neural networks, when properly packaged, allow us to take on these really large, complex analytical cases because they do not depend on the intelligence, insight, or skills level of the person executing them to discover what the patterns are. Neural networks were, in fact, developed as a part of the study into artificial intelligence (or making it possible for computers to think like humans do).

While a wide variety of neural networks are available in the market place today, in general, two types are finding the most commercial acceptance and viability. These networks are used to do two things:

1. Predict customer buying and spending behaviors;
2. Evaluate the relative credit risk that any given individual represents.

While we will concentrate our example on the prediction of customer behavior, be aware that the approach, techniques, and execution sequence for the credit risk assessment problems are basically the same.

10.2 How can a neural network help with marketing?

The basic modeling process is simple and straightforward. A file is constructed that contains all of the information we can possibly find about people who have purchased our products or services in the past. This file is then read by the neural network, which analyzes all of the different characteristics of those people and the amounts of money they spent over a given period of time. When its analysis is finished, the neural network will assign a "weight" or a "score" to each customer. This score tells us how likely that same person is to continue in this same buying behavior based on the patterns exhibited by other buyers with the same characteristics.

While this kind of insight might be interesting and perhaps helpful in predicting that customer's future revenue potential, the usefulness of the analysis does not stop there. Once the neural network has figured out what relative effect each of these characteristics has on the buying behavior, it can apply the knowledge it has learned to new groups of potential customers.

By feeding a collection of information about potential customers, we can get the neural network to tell us which of them are the most (or least) likely to respond to our advertising and offers for service. The net result is that through the use of neural networks, we can significantly reduce the cost of performing marketing activities, while at the same time greatly increase the market share we are able to capture. The neural network can tell us what types of people we should be going after in the first place! Think about it! What better solution to the problems of marketing could you possible ever imagine? By simply feeding a bunch of data into a neural network program, we can save ourselves hours of market research, days or weeks of statistical analysis effort, and millions of dollars in wasted advertising, direct marketing, and direct sales expense. This would be a huge boon to any telecommunications firm.

10.3 Step-by-step use of a neural network

To demonstrate how a neural network product works, we will use another simple case study. In this situation, ABC Telecommunications International has assembled a file with information about its existing customers. Included in the file are fields that convey who the customers are, where they live, birth month and year, the last time they paid a bill, how much they paid, how long they have lived at the same address, and a long list of other characteristics. This file is read into ModelMAX, the neural network tool used in this case, and the variables are displayed on the data edit screen. Figure 10.1 shows a number of the defined variables, along with their physical locations, offsets, and counts.

Once the file is read into the neural network tool, the system reads the data and "learns" about those customers and their characteristics as they relate to our targeted result (increased sales, reduced credit risk, etc.). Letting the network program learn about the population is referred to as "training" the file. Figure 10.2 shows the output of a training run against our sample input telecommunications data set.

10.3.1 What does the training report tell us?

Looking at this training file report, we can get a pretty good idea of what the neural network has figured out about the population. The output of the training run tells us that ModelMAX has evaluated all of the variables and determined that the first 10 variables (from HOW LONG AS CUSTOMER to PAID INDICATOR) all provide valuable, predictive input that can help us ascertain

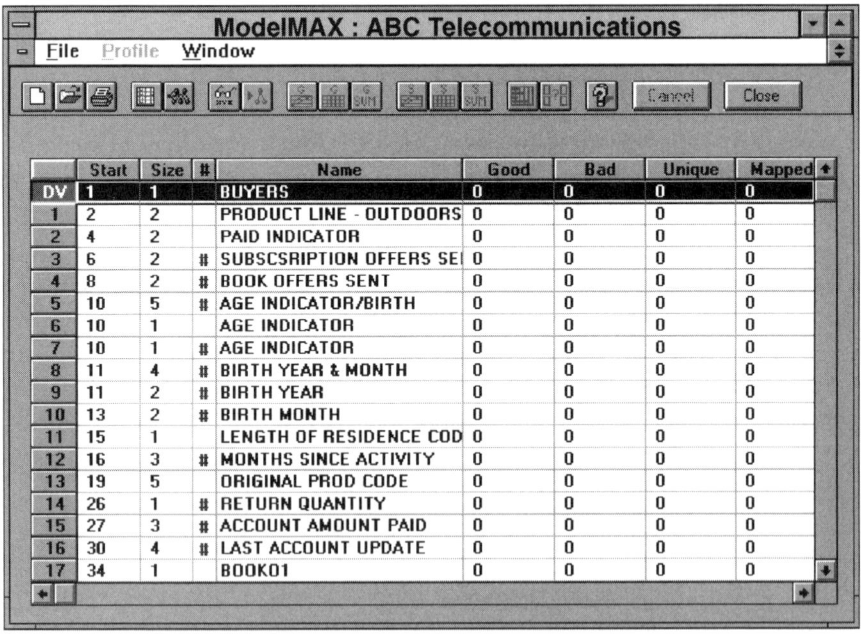

	Start	Size	#	Name	Good	Bad	Unique	Mapped
DV	1	1		BUYERS	0	0	0	0
1	2	2		PRODUCT LINE - OUTDOORS	0	0	0	0
2	4	2		PAID INDICATOR	0	0	0	0
3	6	2	#	SUBSCSRIPTION OFFERS SE|	0	0	0	0
4	8	2	#	BOOK OFFERS SENT	0	0	0	0
5	10	5	#	AGE INDICATOR/BIRTH	0	0	0	0
6	10	1		AGE INDICATOR	0	0	0	0
7	10	1	#	AGE INDICATOR	0	0	0	0
8	11	4	#	BIRTH YEAR & MONTH	0	0	0	0
9	11	2	#	BIRTH YEAR	0	0	0	0
10	13	2	#	BIRTH MONTH	0	0	0	0
11	15	1		LENGTH OF RESIDENCE COD	0	0	0	0
12	16	3	#	MONTHS SINCE ACTIVITY	0	0	0	0
13	19	5		ORIGINAL PROD CODE	0	0	0	0
14	26	1	#	RETURN QUANTITY	0	0	0	0
15	27	3	#	ACCOUNT AMOUNT PAID	0	0	0	0
16	30	4	#	LAST ACCOUNT UPDATE	0	0	0	0
17	34	1		BOOK01	0	0	0	0

Figure 10.1 Data edit screen for a neural network. (Reprinted with permission from Advanced Software Applications.)

```
***Preprocessing started at 09:04:58 AM, Monday, December 18, 1995

Training file is D:\MODELMAX\ABC1.TXT
Record size is 36 bytes (excluding CRLF)
Reading 32162 training records...
Training File:   32162 = (31541 + 621)
Training Sample: 10000 = (5000 + 5000)
Sampling records...
Selecting variables...
R**2       F           ID Name
0.032495   41.680992   14 HOW LONG AS CUSTOMER
0.009373   12.130499   08 BIRTH YEAR & MONTH
0.008093   10.554499   20 BOOK03              = "Paging"
0.014173   18.748896   17 LAST ACCOUNT UPDATE
0.006819    9.079000   07 AGE INDICATOR
0.005267    7.046625   12 MONTHS SINCE ACTIVITY
0.002999    4.022216   11 LENGTH OF RESIDENCE CODE
0.002887    3.880805   01 PRODUCT LINE - = "PCS"
0.002817    2.880805   21 PRODUCT LINE - = "Cellular"
0.001978    2.663182   02 PAID INDICATOR
       *** Suggested Cutoff ***
0.001329    1.789664   19 BOOK02              = "Vmail"
0.001102    1.484992   06 AGE INDICATOR       = "9"
```

Figure 10.2 Training output.

the value of a customer. Notice the *** Suggested Cutoff *** line of the report. The neural network determined that after this point, the variables no longer provide information of any value to the predictive model being developed.

In addition to producing an audit report about what it had learned, Model-MAX also populates several key fields within the data edit screen, as shown in Figure 10.3.

The Good, Bad, Unique, and Mapped fields have now been populated. Good denotes how many usable values were found in the input file. Bad indicates if any values were not as expected (an alpha character in a field indicated as numeric, for example). Unique tells us how many distinct values for that field were found in the file, and Mapped expresses how many bins the neural network will establish for that variable.

10.3.2 Creating and interpreting the gains table

After training the file, we are ready to run a gains report and look at a gains table, as shown in Figure 10.4. These reports tell us how good a job the neural network did of predicting behaviors.

	Start	Size	#	Name	Good	Bad	Unique	Mapped
DV	1	1		BUYERS	32162	0	2	2
1	2	2		PRODUCT LINE - OUTDOORS	32162	0	3	3
2	4	2		PAID INDICATOR	32162	0	2	2
3	6	2	#	SUBSCSRIPTION OFFERS SE	32162	0	11	5
4	8	2	#	BOOK OFFERS SENT	32162	0	2	2
5	10	5	#	AGE INDICATOR/BIRTH	32162	0	144	16
6	10	1		AGE INDICATOR	32162	0	5	5
7	10	1	#	AGE INDICATOR	32162	0	5	5
8	11	4	#	BIRTH YEAR & MONTH	32162	0	252	16
9	11	2	#	BIRTH YEAR	32162	0	84	16
10	13	2	#	BIRTH MONTH	32162	0	13	10
11	15	1		LENGTH OF RESIDENCE COD	32162	0	10	3
12	16	3	#	MONTHS SINCE ACTIVITY	32162	0	176	10
13	19	5		ORIGINAL PROD CODE	32162	0	325	1
14	24	2	#	TIMES IN HEALTH CATEG	32162	0	24	6
15	26	1	#	RETURN QUANTITY	32162	0	10	5
16	27	3	#	ACCOUNT AMOUNT PAID	32162	0	408	16
17	30	4	#	LAST ACCOUNT UPDATE	32162	0	14	10

Figure 10.3 Variables report. (Reprinted with permission from Advanced Software Applications.)

The gains table in Figure 10.4 shows us how the neural network has divided our population of customers by profitability. As we can see, the file has been divided into almost equal-sized segments, each representing approximately 5% of the full population. Another column of importance on this screen is the Response Rate (%). This column tells us how many of the individuals in this category responded (or are most likely to respond) to the next marketing/promotional activity.

At first glance you may be quite surprised at how low that predicted response rate is. Five percent? How can a 5% response rate result in any kind of efficiency in marketing?

Actually, a 5% response rate is remarkable when you consider the "normal" results of untargeted marketing activities, which generally yield a response rate

ModelMAX : ABC Telecommunications

File Window

Cancel Close

			Profit / BUYERS	0.0000					
Seg #	Records	BUYERS	Cost Per Record	Cum. Profit	% of Records	Response Rate (%)	Cum. % Records	Cum. % BUYERS	Cutoff Score
1	1609	82	0.00	0.00	5.00	5.10	5.00	13.20	0.
2	1610	70	0.00	0.00	5.01	4.35	10.01	24.48	0.
3	1603	50	0.00	0.00	4.98	3.12	14.99	32.53	0.
4	1608	51	0.00	0.00	5.00	3.17	19.99	40.74	0.
5	1606	35	0.00	0.00	4.99	2.18	24.99	46.38	0.
6	1620	33	0.00	0.00	5.04	2.04	30.02	51.69	0.
7	1613	38	0.00	0.00	5.02	2.36	35.04	57.81	0.
8	1612	38	0.00	0.00	5.01	2.36	40.05	63.93	0.
9	1608	31	0.00	0.00	5.00	1.93	45.05	68.92	0.
10	1597	31	0.00	0.00	4.97	1.94	50.02	73.91	0.
11	1607	30	0.00	0.00	5.00	1.87	55.01	78.74	0.
12	1623	23	0.00	0.00	5.05	1.42	60.06	82.45	0.
13	1597	20	0.00	0.00	4.97	1.25	65.02	85.67	0.
14	1573	18	0.00	0.00	4.89	1.14	69.91	88.57	0.
15	1612	15	0.00	0.00	5.01	0.93	74.93	90.98	0.

Figure 10.4 Gains table. (Reprinted with permission from Advanced Software Applications.)

of 1% or even less. Identifying a group of people from which I am assured of a 5% return is phenomenal. Of course, the actual predicted response rate is going to vary based on the customers and products being tested. Some models develop response rates in the tens of percentage points. Others may yield minuscule fractions of a percentage points. Remember, there is no guarantee that using an analytical tool will automatically get the kind of results you would like.

Nonetheless, our *best* segment of this file will yield 5.10%. The next highest categories will yield 4.35%, 3.12%, and so forth.

To see how good of a model we actually have, we can take a look at a different output of the system, the gains chart, as shown in Figure 10.5.

10.3.3 Analyzing the gains chart

While the purpose of the gains table is to show us what kind of response rates the different segments should yield, the gains chart reveals how much better than average (or totally random) this predictive model is.

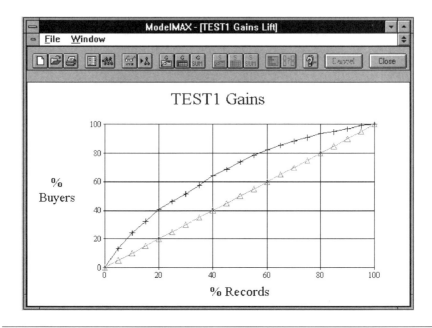

Figure 10.5 Gains chart. (Reprinted with permission from Advanced Software Applications.)

The diagonal line represents the 50/50, or random, probabilities line. If we contacted everyone on our list at random, we could expect to see a one-for-one match between the percent of buyers and the percent of records within the file. In other words, if a file has 1,000 records and there were 10 buyers, we would expect, if we were using a totally random access technique, that every 100 records would turn up another buyer—100 records = 10% of the file (1,000/100). One buyer = 10% (10/1) of the buyers.

The curved line, moving away from the 50/50 line and back again is known as the gains plot. This line shows us how much we can improve the response rate by using the scoring mechanism the neural network has provided. The gains chart for our example indicates that after selecting 20% of the file, we will have already identified 40% of the buyers, and at 60% of the file, we will have 80% of the buyers identified. This shows the improvement over a totally random campaign. The more the line curves upward, the bigger and better the improvement over a random access. The curved line is said to show the *lift*, or improvement over random, that the model guarantees.

Knowing where the cutoff points are allows us to identify those segments of the market that are most economical to address. For example, say that a cellular phone company figures out that it costs an average of $55 to bring on a new customer, and that the potential market they might pursue covers one million people. It is unlikely that they will perform an all-out marketing blitz to acquire all of those customers (for a whopping grand total of $55 million in expense.) A neural network model allows them to target a much smaller group of people who are much more likely to sign up for the services. This means that they get more new subscribers for less up-front cost.

10.3.4 Making marketing programs as profitable as possible

Knowing what the likelihood is of a customer purchasing a certain level of products or services is good information to have. What would really be nice, though, would be to know who we should be marketing to for maximum profit. ModelMAX comes with this capability.

The ModelMAX product allows you to input the cost to acquire a customer into the model. Once you key in the value into the Cost Per Record column, the system can compute the cumulative recognizable profit for each of the identified segments, as shown in Figure 10.6. With this information, it is then a trivial matter to figure out how big or small a marketing campaign should be.

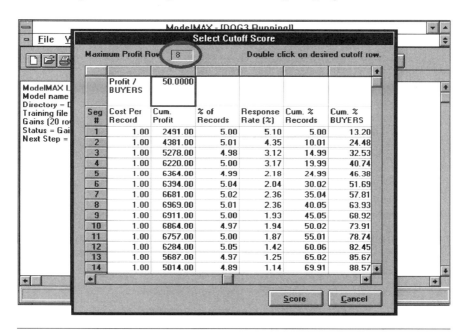

Figure 10.6 Computing cumulative profit/depth. (Reprinted with permission from Advanced Software Applications.)

(To simplify the example and to protect the identity and "real ratios" that telecommunications firms are actually getting, we have used a value of $1.00 as the cost of acquiring a new customer and have deflated the potential revenue values.)

Instead of guessing at how many people you should be addressing through direct sales, direct advertising, and other means, you can simple look to the model and keep going after more customers until there is no more profit to be made.

Notice the box at the top of the Select Cutoff Score screen in Figure 10.6. Here, ModelMAX declares row 8 to be the *Maximum Profit Row*. If you take a close look at the Cumulative Profit column, you will notice that profits start at $2,491 for the first segment, jump up to $4,381 in the second segment, and continue to inch upward until the ninth segment. Row 8 shows a profit of $6,969, and row 9 shows $6,911, a drop of $58. This is the point at which we start to lose some of the profit we will have gained by marketing to the best segments first.

10.4 Applying the model to prospects

Of course, up until now, our entire neural network analysis has been concentrated on analyzing an existing collection of customer data. This analysis, in and of itself, is at best of marginal value. However, the ModelMAX product allows us to address this issue also.

Once our model has been developed and analyzed and we are sure that we like the results this model generates, we are ready to really make things happen. At this time we can take this now intelligent neural network system and allow it to read through prospective customer files. The neural network will be able to go through these lists of prospects and, based on what it has learned during its training run, it will be able to tell us what the potential values of our prospects are. This is known as the scoring process.

After scoring a new file, we can again do the profitability analysis and then target only those prospects who have the greatest likelihood of turning into satisfied customers.

10.5 Conclusion on neural networks

The neural network product considered here is the last of the quantitatively based data mining tools we are discussing in depth. Neural networks represent the latest and most exciting of a long line of innovations in software that are making data warehouses more viable and more powerful than ever before.

Chapter 11

Engineering and competitive analysis support: an introduction to geographical systems and MapInfo

While the data mining methods we have looked at thus far take a decidedly data and reporting approach to empowering telecommunications executives with the information they need to make good decisions, the technique we will be considering next takes an emphatically nontabular bent. While data reported in tabular form is undeniably the more prolific and useful of reporting formats, there is a large class of business problem resolution that this form fails to adequately support.

Geographical information systems (GIS) and, more generally, graphical reporting systems (GRS) enable the warehouse user to solve problems and gain insights into situations that tabular data can only hint at.

The telecommunications industry is highly dependent on a number of important characteristics, with geography, the physical locations of items in relation to other items, being one of the major ones.

11.1 An introduction to MapInfo Professional

MapInfo Professional is one of the oldest and most mature of the geographical information system software packages. Over the years, the developers of this product have assembled a powerful collection of combined data, database, and geographical display technologies into a comprehensive, easy-to-use, highly effective package.

The fundamental principles behind MapInfo Professional are simple and straightforward. The MapInfo Professional environment is designed to display the geographical data requested and blend that information with the business-related performance data that assists users in the analysis of their business. This combined display of graphical and tabular data then provides the user with a dynamic and functional view of what the current business environment really looks like.

It should come as no surprise that the developers of MapInfo Professional have dedicated a significant amount of their efforts to the creation of a lot of capabilities specific to telecommunications companies. Their product is used by these firms all over the world to solve hundreds of previously practically unsolvable problems.

11.1.1 MapInfo telecommunications offerings

In addition to the "traditional" features that any geographical information system will offer—namely, the ability to display maps, latitude/longitude, streets, waterways, and a plethora of other standard business-related geographics of interest—you can also purchase specific overlays with a decidedly telecommunications emphasis. These specialized offerings include the following (we include here only the U.S. telecommunications-specific offerings for the sake of saving space; other offerings for other geographical areas are also available):

- LATAInfo™ (a map database from On Target Mapping)—Map database of local access and transport area (LATA) boundaries. LATAs define the geographical areas through which telephone traffic may pass. LATA boundaries define the way telecommunications companies bill each other for commonly carried traffic;

- StreetWorks™ (MapInfo Corp.)—Nationwide detailed street maps that can be tied to customer data. Using StreetWorks, you can map custom-

Engineering and competitive analysis support   157

ers, prospects, trouble spots, and a variety of other people and events to a detailed plan, showing their precise location;

- ExchangeInfo™ (a map database from On Target Mapping)—A database containing wire center serving areas, which are the basic unit of geography in the telecommunications industry (pinpointing which exchanges cover which area);
- CellularInfo™ (a map database from On Target Mapping)—A mapping database of cellular coverage in every cellular market in the United States (including market side, station licensee, signal threshold, transmitter power, antenna height, gain, and number of antennas);
- Trendline-GIS—Displays past, present, and future census data for specific geographical areas;
- MapMarker™ (MapInfo Corp.)—Ties tabularly recorded addresses (customers, troublespots, etc.) to physical locations on the map (referred to as the geocoding process);
- LECInfo™ (a map database from On Target Mapping)—Maps the unique service territories of all local exchange service companies in the United States;
- ExchangeInfo Plus™ (a map database from On Target Mapping)—A database of wire center serving areas;
- ContourInfo™ (a map database from On Target Mapping)—A nationwide contour elevation data file, making it possible to determine line-of-sight and line-of-scope capabilities for transmitters and repeaters;
- CableInfo™ (a map database from On Target Mapping)—Comprehensive information on cable systems and their boundaries;
- TrafficVolumes—The 24-hour average daily traffic counts for BLR major roads;
- MSA/RSAInfo™ (a map database from On Target Mapping)—Monthly or quarterly updated reports on metropolitan and rural service areas, allowing users to see their markets and pertinent information about each area;
- AreaCodeInfo™ (a map database from On Target Mapping)—Maps the boundaries of all of the nation's three-digit area codes;
- PCSInfo™ (a map database from On Target Mapping)—The method for identifying collections of counties and towns around major metropolitan areas or cities.

With so many different products and offerings that specifically target the needs of telecommunications firms, it is difficult to decide which ones to use as typical examples for our coverage here.

11.2 Using geographical information to solve telecommunications problems

We will start by considering some of the many ways that this kind of information can be turned into useful insights by the savvy telecommunications user:

- *Opportunity analysis*—You may recall that among our list of different business functions that might be part of a telecommunications firm's value chain was the process of identifying the places where the services should be offered and acquiring the rights to service those areas. Geographically displayed information can be an incredible boon to the person responsible for making this kind of analysis. The ability to see what kind of coverage is being provided by competitors, when combined with census information about the economic potentials those areas represent, can help the analyst make better informed decisions.
- *Customer service*—Some organizations have created customer service systems that include the mapping of customer locations to physical maps. When customers call in, the customer service rep is able to immediately pinpoint their physical location. At the same time, service vehicles are tracked using global positioning systems (GPS) and their current locations are fed into the same mapping system. This allows customer service reps to dispatch resources as efficiently and timely as possible, greatly increasing customer satisfaction while reducing cost at the same time.
- *Marketing and sales analysis*—By overlaying geographical, customer, demographic, and census information, marketers and sales people are able to identify the best market areas to go after.
- *Engineering (wireless and fiber)*—The strategic combination of geographical, customer, and market share information, when overlaid with the physical composition of the firm's own and of competitors' distribution of resources, allows the engineer to identify coverage overlap and weaknesses in competitors' configurations, and makes it possible for that engineer to identify strategic opportunities for maximum configuration enhancement.

11.3 Cellsite analysis with MapInfo Professional

As we already mentioned, the makers of MapInfo Professional have developed a diverse collection of databases and support products to assist the telecommunications decision makers with challenges specific to their industry. One of the major areas of support coverage is that of cellular phone service.

Users of the system can acquire geographical mappings of the areas within which their services are offered and perform all kinds of analysis based on the positioning of their cellular sites. Figure 11.1 shows the mapping for a relatively small geographical area.

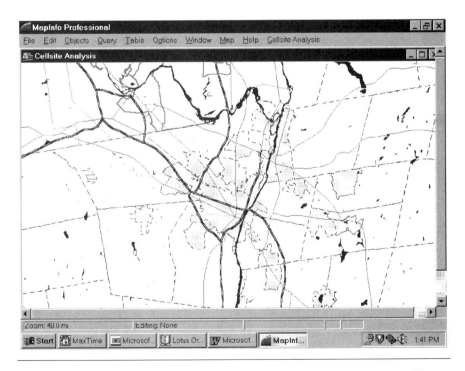

Figure 11.1 A geographical area display. (Maps made with MapInfo Professional™, MapInfo Corporation, Troy, New York.)

This view shows us the major residential areas in this region (highlighted with the faded checker pattern at the center of the screen), along with major and minor roadways (the heavy and thin black solid lines that traverse the screen), the location of rivers and small lakes (black splotchy areas), and geographical boundaries (the dotted lines).

After identifying the geographical area we want to analyze, we can simply go to the MapInfo Professional menu bar and choose "Cellsite Analysis." (See Figure 11.2.)

The Cellsite Analysis menu offers many options, including Load Maps [lets us overlay the location of cellular sites or the location of previously reported failed cellular service events (the Tech Support Logs option)], Generate ReUse Grid, Locate Problem Area, Run Signal Analysis, and others. We will choose the Cellsite Map and Tech Support Logs from the Load Maps option to provide you with a general idea of how the process works. The results of these selections can be seen in Figure 11.3.

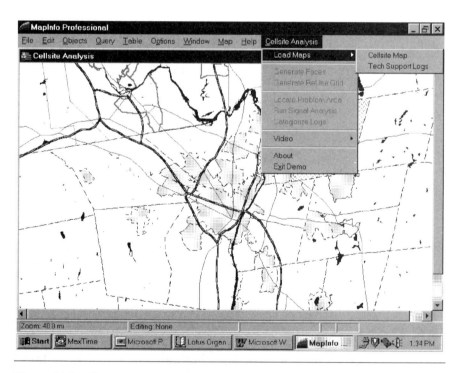

Figure 11.2 Overlaying the map with cellular data. (Maps made with MapInfo Professional™, MapInfo Corporation, Troy, New York.)

Engineering and competitive analysis support 161

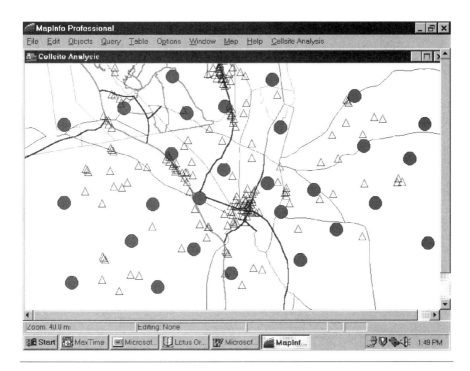

Figure 11.3 Cell sites (circles) and trouble reports (triangles). (Maps made with MapInfo Professional™, MapInfo Corporation, Troy, New York.)

Figure 11.3 shows us the same map we were looking at earlier, except that many of the noncritical details (geographical boundaries and waterways) have been removed. Instead, what we see are an assortment of large dark circles (which represent the locations of different cellular sites) and triangles (which tell us where cellular services have failed in the past).

By displaying this information in this graphical form, it becomes a trivial matter for the engineer to figure out whether trouble reports are random events or part of a bigger overriding pattern. More importantly, the engineer can actually notice those areas where a significant number of trouble spots occur. This makes if possible for him or her to investigate further and consider the possibilities of some additional problems in particular calling areas.

In this case, for example, we might decide that we need to have some more detail about that large area of trouble calls near the center of the screen. MapInfo Professional allows us to zoom in and zoom out at any level we happen to be looking at the map. So, the engineer only needs to zoom in on this section to see if more detail will give a better indication of what the real problem is.

Figure 11.4 shows us the zoomed-in view of this troubled area. Notice the relative positioning of cellular sites in relation to the troubled areas. Clearly, either the repair or enhancement of less than optimal equipment, or the placement of some additional equipment, is called for if this troubled area is going to be addressed effectively.

11.4 Market analysis capabilities

MapInfo Professional offers more than just simple cellsite analysis capabilities. In the next example, we will take a look at a much more robust application. This application, called the Telecom Mapper, bundles many of the MapInfo telecommunications databases and offerings into one comprehensive analysis package.

We will start by looking at the way MapInfo Professional allows an analyst to gain different perspectives on the company's relative competitive strength.

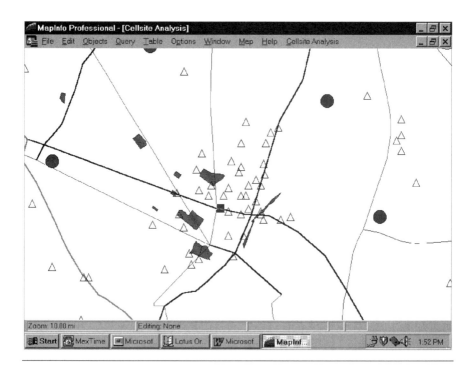

Figure 11.4 Zooming in on trouble reports. (Maps made with MapInfo Professional™, MapInfo Corporation, Troy, New York.)

The first screen we see (Figure 11.5) shows a view of the entire United States, divided into the seven major Regional Bell service areas.

This mapping shows the boundaries between each of these areas. From this screen we can change the view, adding details of interest to either Long Distance, Local Market, or Cellular-PCS users, by invoking the options within the corresponding menus.

For example, the Long Distance menu allows users to view the locations of switches, the links between those switches, the boundaries between different exchanges, LATAs, or area codes, and provides the option to add cities, highways, county boundaries, or a weather map overlay. In Figure 11.6, we have selected a map of South America and the Switch Locations option.

Besides all of the viewpoint options, the Long Distance Analysis system allows us to locate specific exchanges, shade the links, view the links for selected switches or by range of values, or to actually monitor link activity.

As is true with most MapInfo views, the product allows us to perform this analysis not only at the very high level view, but at any lower level as well. We

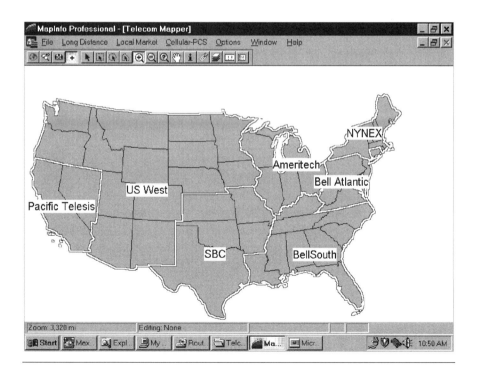

Figure 11.5 U.S. market areas. (Maps made with MapInfo Professional™, MapInfo Corporation, Troy, New York.)

Figure 11.6 Long-distance analysis options: switch locations, South America. (Maps made with MapInfo Professional™, MapInfo Corporation, Troy, New York.)

can choose to zoom in on one region of the country, one state, one city, or simply one exchange area, pulling in and removing different overlays and perspectives as needed.

11.5 Viewing a local market in greater detail

Assuming that we have zoomed down to a specific area of the United States, in this case the southern tip of Lake Michigan (Indiana and Illinois), we get a chance to see a little more of what MapInfo Professional can offer. Figure 11.7 shows this area with boundaries marked out by state (solid black line), basic trading area (BTA, marked by the dashed line), and metropolitan trading area (MTA, showing the Chicago metropolitan area in the shaded region).

Choosing the Locate Exchange Boundary option from the Local Market drop-down menu brings up an exchange selection box, and we are asked to enter the area code, the first three digits of the phone number, and whether or not we want demographic information included in the display (see Figure 11.8).

Engineering and competitive analysis support

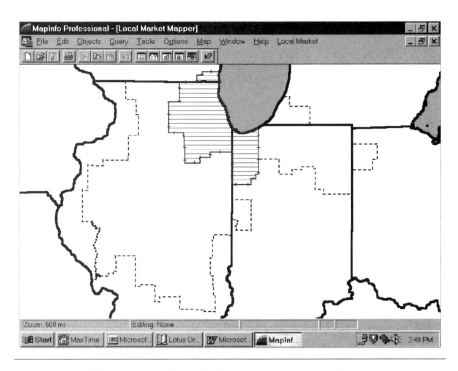

Figure 11.7 Chicago BTA, MTA, and adjoining areas. (Maps made with MapInfo Professional™, MapInfo Corporation, Troy, New York. *Source*: PCSInfo™, On Target Mapping™, 1997. All Rights Reserved.)

The results of this will be the map shown in Figure 11.9.

The resultant map automatically drills us down to the sector of the city that carries that area code and three-digit exchange. While we could now ask the system to overlay roadway, waterway, cellular site, fiber-optics channel, or any of a number of different types of information, at this point we will be most interested in the simple demographics of the area. Notice the legend box in the upper left hand corner of the screen. This legend tells us the average income for each household (HH) in the area, as well as the number of homes that each income level relates to.

11.6 Accessibility to fiber analysis

One last example will help us to finalize our review of the MapInfo product. We will start with our same Local Market Mapper perspective, but this time we will

166   DATA WAREHOUSING AND DATA MINING FOR TELECOMMUNICATIONS

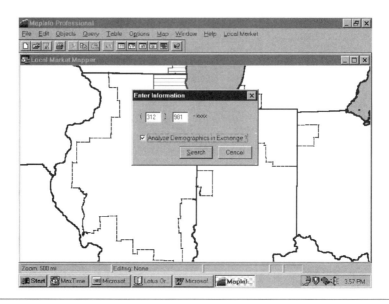

Figure 11.8   Choosing to locate an exchange. (Maps made with MapInfo Professional™, MapInfo Corporation, Troy, New York. *Source*: Claritas Inc., Arlington, Virginia. *Source*: ExchangeInfo Plus™, On Target Mapping™, 1997. All Rights Reserved.)

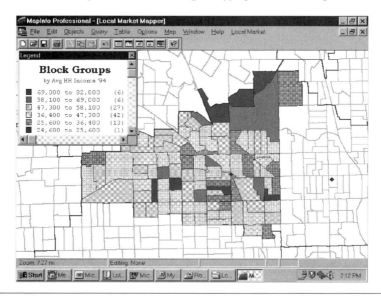

Figure 11.9   Demographics for a selected exchange area. (Maps made with MapInfo Professional™, MapInfo Corporation, Troy, New York. *Source*: Claritas Inc., Arlington, Virginia. *Source*: ExchangeInfo Plus™, On Target Mapping™, 1997. All Rights Reserved.)

be using MapInfo Professional to help us analyze some of the capabilities that our existing network may or may not be able to support.

After choosing the area code and exchange that we would like to analyze in greater detail, we will select Analyze Area in Fiber Limits from the Local Market menu options. The product runs the analysis and then shows us an up-close view of the actual buildings located in the area.

To see what the potential "reach" would be, we can choose the Recalculate Fiber Limits option and a selection box will let us change the default Actual Loop Limit of 500 to 1,000 or whatever other level we would like to reset it to. The result of this will be the display of a map showing the original and extended ranges, as shown in Figure 11.10.

11.7 Working with the underlying database

Not only does MapInfo allow you to manipulate the graphically and geographically based information, it also comes equipped with a database query and

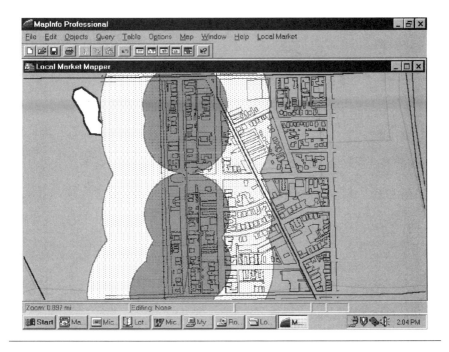

Figure 11.10 Viewing the area for fiber accessibility. (Maps made with MapInfo Professional™, MapInfo Corporation, Troy, New York.)

search tool that allows you to work with and report on the underlying data in a tabular format, as shown in Figure 11.11.

This screen shows the Query and Select capabilities screen invoked. A user can build any kind of query to go against any of the underlying tables that make up the MapInfo space.

11.8 Conclusion

By allowing marketers, engineers, sales people, customer support personnel, and anyone else to combine this ability to display data in a graphical format with the ability to zoom in and out of different areas of interest, MapInfo Professional becomes a required toolset for many telecommunications professionals. Given a well-designed data warehouse and a couple of tools like this one, it is easy to see how a telecommunications firm can gain some serious competitive advantage while decreasing costs at the same time.

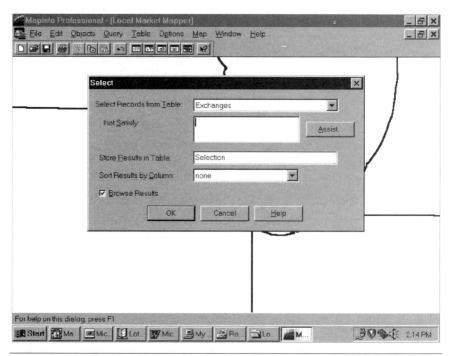

Figure 11.11 Working with the underlying databases. (Maps made with MapInfo Professional™, MapInfo Corporation, Troy, New York. *Source*: ExchangeInfo Plus™, On Target Mapping™, 1997. All Rights Reserved.)

Appendix A

Real world warehousing: France Telecom and STATlab tools

by Michel Jambu, France Telecom-CNET
and Jean Schmitt, SLP InfoWare Inc.

THE WORLDWIDE telecommunications market is rapidly transforming itself from a monopolistic industry to one that is fiercely competitive and dynamic. The key factor for this change is complete deregulation in the European Union by January 1, 1998, the same process that started in the United States several years ago.

Competition makes data more important than ever before. For example, customer surveys have found that in a monopolistic setting, only 10% of the typical telecommunications company's customers were dissatisfied with that company's service. What is the economic penalty if the monopolistic company decides to do nothing about that 10%? The answer: none.

In a competitive situation, however, the challenge is not only to be good, but to be better than one's competitors. This simplified scenario implies concepts that can impact the way companies conduct their business:

- Information is not necessarily the same in a monopolistic situation;
- Information is viewed as a necessary step in the decision-making process;
- Information on competitors and their customers is vital;
- In a monopolistic situation, the information system is an internal management tool;
- In a competitive situation, the information system is a customer's management tool.

The challenge then is to have the right data accessible and analyzed by the right data tools, manipulated by the right decision makers. At France Telecom we worked to meet that challenge.

A data warehouse solution with STATlab tools

In 1990, researchers from the Center for Telecommunications Research of France Telecom (CNET) and SLP InfoWare Inc., formed a partnership to design a new generation of data analysis and information system tools. The challenge was to design a unique enduser data workstation that made all the different kinds of data access and analysis tools available, and be easily connected to data servers which in turn could be connected to any part of any organization's overall corporate information systems structure.

Out of this partnership came the *STATlab Data Warehouse Solution* (a user data warehouse with STATlab tools).

We began by compiling a comprehensive collection of the "work methods" of the decision makers. Then the basic core, entitled STATlab, was designed as a data analyzer for any decision maker. It was completed with satellite software such as data briefing and reporting modules, mapping tools for quality control data, statistical quality control facilities, and many others. Finally, the data server and the associated data query tools were built. These were adapted to enable users to manipulate data and access the corporate data warehouse environment.

As is often the case, users end up needing to invoke several tools to do their daily data analysis work; a tool for the database, a tool for query or data access, and several tools for analysis and presentation. Our challenge was to provide a unique, coordinated toolset and make it available regardless of the computer hardware and network. As an added bonus we also tried to design it in a way that would make the end users comfortable and give business advantage to their enterprise. The STATlab solution made that possible at France Telecom.

What makes the STATlab solution different from other possible or existing data warehouse projects is that it was developed *with* and *for* decision-makers and managers, the data warehouse users, and not imposed upon them by vendors. The STATlab solution is now available for distribution to any organization.

The objective of any user data warehouse solution should be to create an environment that allows users to do their daily work without having to wrestle with the computer environment and the data analysis process, no matter what kind of data is involved, or what kind of hardware/software their company uses. The STATlab solution allows any enduser to work with a minimum of these kinds of problems.

The examples cited throughout this chapter are from France Telecom, the French telecommunications operator. But similar types of data, business situations, and management challenges apply to any other telecommunications operator.

The corporate information system of France Telecom

The France Telecom information system involves many projects and computer applications. It is divided into four main domains:

1. The *commercial domain* (with computer applications);
2. The *network domain* (with over 200 computer applications);
3. The *human resources domain* (with approximately 30 computer applications);
4. The *finances and management domain* (with approximately 100 computer applications).

These applications communicate with each other, which can make things very difficult to manage. To give the reader an idea of the importance of the internal information system at France Telecom, there are about 160,000 agents using about 109,700 workstations (PCs and terminals) and 25,000 printers. These statistics (given for 1992) illustrate the critical importance of data for telecommunications operators.

In addition to the four domains, there are information systems not part of the internal structure:

1. *The customer surveys and marketing information system.* This type of information is provided by the consumers or clients of France Telecom. It includes information that is economical, sociological or demographic in nature and may also be provided by public institutions. Of course, the France Telecom decision-maker must manipulate both internal and external information to be successful at analyzing the business effectively.

2. *Competitive information.* This is relatively new information for telecommunications operators that comes from different sources. For example:

    - Information related to technological monitoring or the registration of telecommunications patents;
    - Information related to the detailed budgets given for specific research or technology (e.g., the information circulating through journals, magazines, exhibitions, and conferences on a specific telecommunications subject);
    - Information related to the indicators of performance of telecommunications operators that allows them to compare each other according to a set base of metrics.

3. *The employee management information system.* More and more, employees play an important role in competitive analysis because the quality of service provided has a big impact on customer satisfaction. Making good human resources decisions plays an important role.

Cases of telecommunications data exploration

Visualization of the telephone network's quality of service data

A computer application, named 39A, was built to manage the telephone network quality of service. This application gives managers performance indicators about network reliability, and is refreshed monthly. Users' data workstations have STATlab tools (data mining tools from SLP InfoWare Inc., detailed later in this chapter) and Microsoft Excel as the user interface software.

Data comes from two principle sources. One is a record of all telephone traffic for a given time period, and the other tracks occurrences of clients' out-of-order signals. Geographic and technical dimensions are associated with this data (the switching units are known by their type and model). This means the user can study the statistics related to traffic and out-of-order signals from switching units to the top level. There are as many data users (approximately 350) as there are decision-makers (from the switching exploitation center to the executive).

The data server is a Unix machine that records all data coming from the switching units center for a 48-month period. The data are online and available to every data user.

The principle data access approach provided is very simple and was designed based on decision-makers' requests. (These users were reluctant to use the SQL language. For them, it was important to have access to the variables, entities, periods, and other data without needing to learn a programming language.). See Figure A.1 for the initial entry screen for this system. As the figure shows, users selected a total of 9 entities (left), 12 variables (middle), and

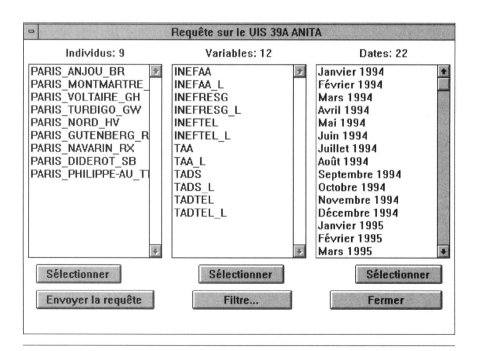

Figure A.1 Display of a data request. (Reprinted with permission from CNET-France Telecom and SLP InfoWare Inc.)

22 periods (right). After sending the request to the data server, it is automatically filed within STATlab. The data is then ready to be used for exploring and visualizing by the decision makers. (Since the examples are from France Telecom, all screen shots are in French.)

The data request can be transferred to a spreadsheet like Excel to run a data presentation program. The process for accomplishing this is as follows: First, a data population is selected using the data directory. (See Figure A.2.) Then, the data to be extracted are sent to an Excel spreadsheet. (See Figure A.3.) But a better way is to use STATlab as a data analyzer. (See Figure A.4.)

In addition to the basic reporting capability the user can also do more sophisticated processing. Figure A.5 shows a simple boxplot and dotplot.

Several possibilities are given to help users view chronological data. Among them, there are selected successive boxplots (Figure A.6), dotplots, and timeplots where each line represents a different entity.

Figure A.7 shows a time series report, with each line tracking the values of different entities for the specified time periods.

You can also use the system to view data geographically. Figure A.8 shows represented data on a map.

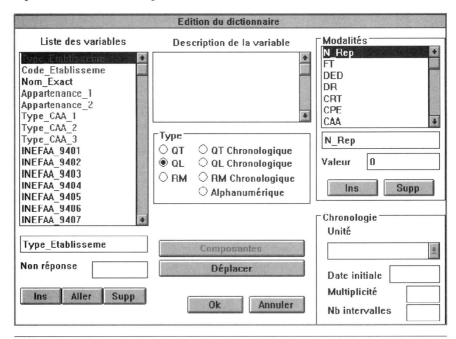

Figure A.2 The current data directory. (Reprinted with permission from CNET-France Telecom and SLP InfoWare Inc.)

Appendix A: Real world warehousing 175

	1	2	3	4	5	6	7	8	9
1	Label	SINUM_9510	SILLUPT_9510	SITFZ_9510	INCIDENT_9510				
2	PARIS_NORD_U5	24.10	9.50	0.10	0.30				
3	PARIS_ANJOU_BR	20.30	9.90	0.20	0.20				
4	PARIS_MONTMARTRE	27.70	8.00	0.10	0.20				
5	PARIS_VOLTAIRE_GH	25.00	11.00	0.20	0.20				
6	PARIS_TURBIGO_GW	25.50	8.00	0.10	0.10				
7	PARIS_NORD_HV	26.80	10.10	0.10	0.60				
8	PARIS_GUTENBERG_R	19.20	8.50	0.20	0.20				
9	PARIS_NAVARIN_RX	22.60	10.60	0.10	0.20				
10	PARIS_DIDEROT_SB	26.20	10.90	0.10	0.20				
11	PARIS_PHILIPPE-AU_T	27.90	7.60	0.20	0.30				
12									

Figure A.3 Data export from STATlab to Excel. (Reprinted with permission from Microsoft Corporation.)

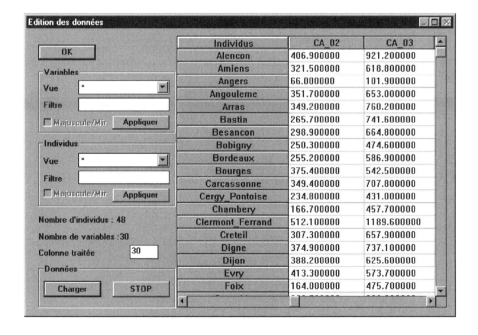

Figure A.4 The STATlab spreadsheet. (Reprinted with permission from CNET-France Telecom and SLP InfoWare Inc.)

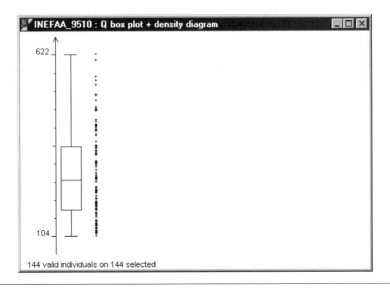

Figure A.5 Simple boxplot and dotplot. (Reprinted with permission from CNET-France Telecom and SLP InfoWare Inc.)

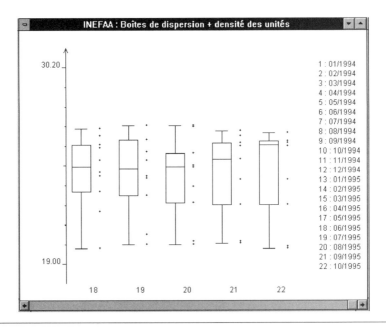

Figure A.6 Boxplots for a series of time periods. (Reprinted with permission from CNET-France Telecom and SLP InfoWare Inc.)

Appendix A: Real world warehousing 177

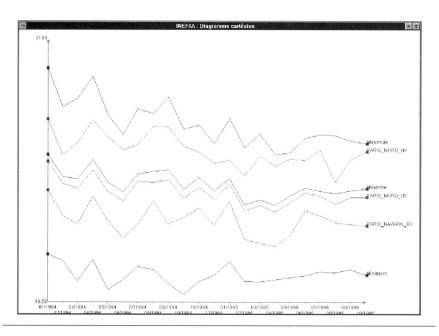

Figure A.7 Time series report. (Reprinted with permission from CNET-France Telecom and SLP InfoWare Inc.)

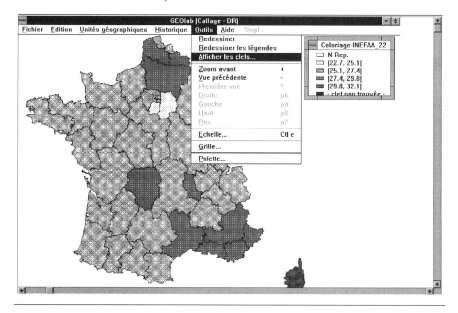

Figure A.8 Geographic view of data. (Reprinted with permission from CNET-France Telecom and SLP InfoWare Inc.)

A special data report can be prepared by the software associated with the STATlab system. It is a dynamic reporter, meaning that reports can be updated automatically as soon as data are received. Figure A.9 shows an example of the data reporter's output.

Visualization of consumer quality of service data

A quality of service information system gives decision-makers or managers useful information on a timely basis that helps them to plan and take action to improve the quality of service to their clients. In large organizations such as France Telecom, this information must be given to managers at every management level in every branch, department, or market segment.

There are at least two types of quality of service data. One is information gained by measuring either the performance of the telephone network itself, or the behavior, characteristics, or responses of consumers (time lag until a customer is initially connected to the network, response time for operator services, postdialing delay, successfully established call attempt rate, speech transmis-

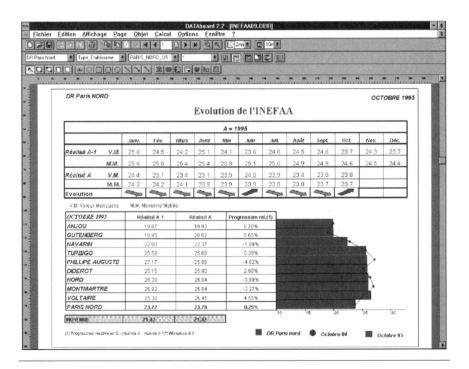

Figure A.9 DATAboard output. (Reprinted with permission from CNET-France Telecom and SLP InfoWare Inc.)

sion quality, faults reports per access line per year, fault repair time, availability of public pay phones, billing complaint rate, or dial-tone delay). These data are considered objective or unbiased. The other type of quality of service data is a substantial amount of subjective or perceived data collected from the clients themselves through complaint letters or customer satisfaction surveys.

At France Telecom, a large number of surveys are regularly executed to obtain client opinions of services provided. This information is collected by business units (local business agencies), on the France Telecom network or through other means of communications. About 1,000,000 organizations are polled each year for France Telecom, using a computed aided telephone interviewing (CATI) system.

Data are recorded monthly or quarterly and satisfaction indicators are reported for each level of management at France Telecom. The detailed data (the responses to the questionnaires) are also available to all management-level individuals for their own quality of service business concerns. Questionnaire production is tightly supervised because part of each France Telecom agent's salary depends on the value of the quality of service indicator. A special committee supervises the surveys operation.

There are about 50 questions on the questionnaires, and each question has five possible responses. In addition, some extra information is given by the clients about sex, age, profession, level of billing, and so on. The results are aggregated at the level of each business unit or region; and quality of service indicators are then computed and used by executive management.

The STATlab system is used both for data reporting and data exploitation. Figure A.10 shows the quality of service by business unit report.

Figure A.11 shows a categorical factor analysis report for these same questionnaires.

Visualization of responses to open-ended questions

Another type of survey conducted by France Telecom solicits free-form responses from customers to questions asked by an interviewer. Textual data mining can be used on this kind of raw data and gives managers useful information on the quality of service as perceived by the clients.

There are two types of data accessibility. The first gives managers or decision-makers the verbatim content of the consumer's response. A data server contains all the textual data, which are then disseminated to everyone through the corporate network (according to key access). See Figure A.12.

The second type of data accessibility gives managers true statistical analysis of the free-form text. We will site several examples, but first let us look at a

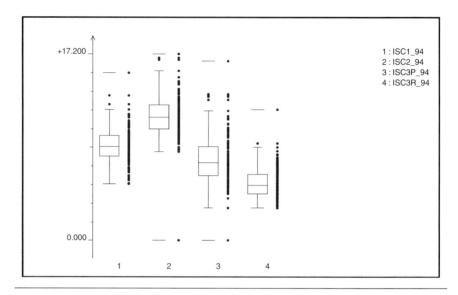

Figure A.10 Quality of service by business unit report.

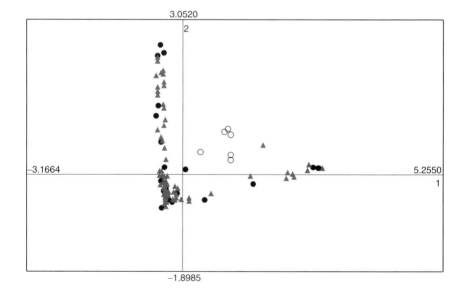

Figure A.11 Categorical factor analysis report.

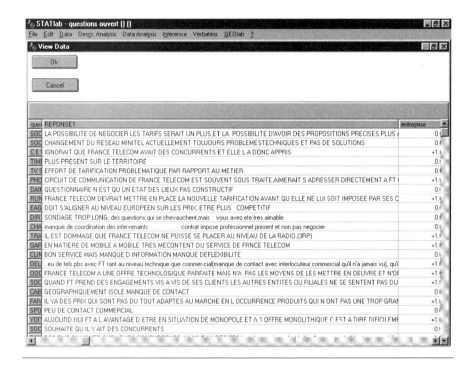

Figure A.12 Textual data of France Telecom customers' survey response. (Note: The responses are given in French, but the method and software associated are applicable in any language.) (Reprinted with permission from CNET-France Telecom and SLP InfoWare Inc.)

sample of a lemmatization (list of words used) and statistics on the words in Figure A.13. Then different types of statistical analysis are made. One extremely useful way of analyzing this information is through the use of a correspondence/factor analysis. This technique tells us, for instance, how different customers' complaints are related to each other. Figure A.14 has a biplot showing correspondence analysis of the data.

On the same biplot, consumers and words used can be seen and the relationships between words and clients can be highlighted (see Figure A.15).

Visualization of marketing data

The database contains approximately 150 corporate variables coming from the corporate information system, and approximately 700 demographic, sociologic, and economic variables drawn from official sources like the French census. The dimensions associated with these data are geographic, (i.e., re-

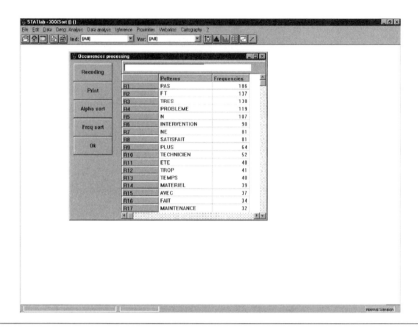

Figure A.13 Frequencies associated with words. (Reprinted with permission from CNET-France Telecom and SLP InfoWare Inc.)

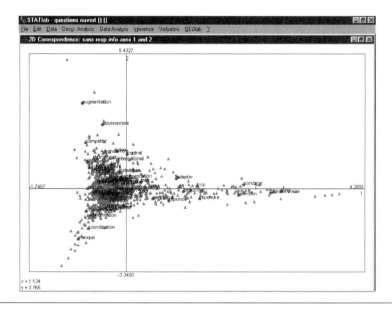

Figure A.14 Biplot of correspondence analysis of words and consumers (axes 1 and 2). (Reprinted with permission from CNET-France Telecom and SLP InfoWare Inc.)

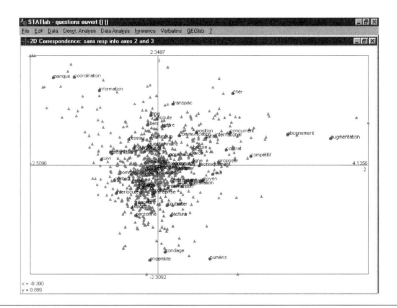

Figure A.15 Biplot of correspondence analysis of words and consumers (axes 1 and 3). (Reprinted with permission from CNET-France Telecom and SLP InfoWare Inc.)

gions, business units, and countries). These dimensions allow business unit managers to do local marketing and business evaluations.

The system STATlab is available for all business managers who have ready access to the data they need and who perform analysis remotely or at their data workstations. This type of marketing information system is available to various departments of France Telecom.

Visualization of modeling, and forecasting the telephone traffic data

The most important data for a telecommunications operator are telephone traffic data. Whenever a new service is introduced, there is a major battle related to new traffic patterns and problems. Visualization of this traffic is the subject of our next exploration of the France Telecom applications area. There are several reasons to make this type of data available for analysis. First, people need daily access to traffic statistics by market-segment in order to forecast pattern changes. In general these data are available for any market segment, switching unit, or traffic recorder centers.

The applicable data tools are STATlab for exploring data, GEOlab for representing statistics on maps, and DATAboard for printing and reporting data. These tools are particularly efficient in highlighting the main features of the telephone traffic data. However, the most interesting tool for traffic forecasting is TIMElab. For example, the traffic recording center (Center d'enregistrement) provides service times for international, national, and local market segments. The variables used are "number of calls" and "number of telephone units." Figure A.16 shows the results for the international segment.

After modeling the traffic, we can forecast it using the Holt-Winters method, which gives the results seen in Figure A.17.

Visualization of billing data

A study was conducted to see if STATlab could visualize billing data directly without computing aggregated data. Billing data are the key decision-making source for much of the business side of a telecommunications firm, so the ability to view this data directly without aggregation would mean that executives and users could see in real time exactly how well the different areas of the business were running.

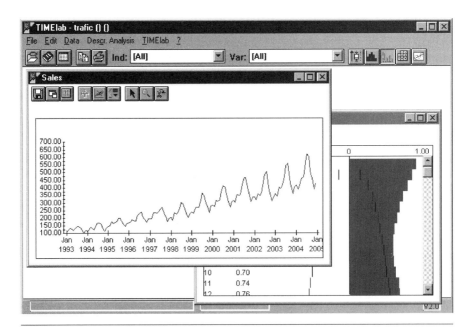

Figure A.16 Residual associated with the regression analysis. (Reprinted with permission from CNET-France Telecom and SLP InfoWare Inc.)

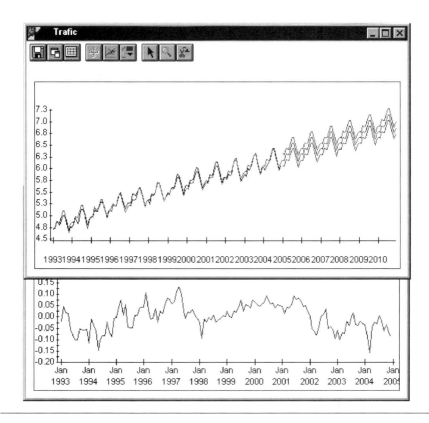

Figure A.17 Telephone traffic forecasting. (Reprinted with permission from CNET-France Telecom and SLP InfoWare Inc.)

The bills are recorded at each billing center. Each bill provides different types of data: traffic, services, rental subscription, and so on. The bills can be viewed according to market segment, region, and other dimensions.

STATlab can easily explore billing data. Figure A.18 shows billing information displayed using a boxplot diagram. This kind of report can help executives learn what the "spreads" are for different variables across the identified population area. The polar diagram in Figure A.19 shows how this same information distributes in relation to other values.

Data reporting for executive managers

It is important for executives to have access to the metrics that determine how well their area is doing in comparison with their key objectives. Once a month

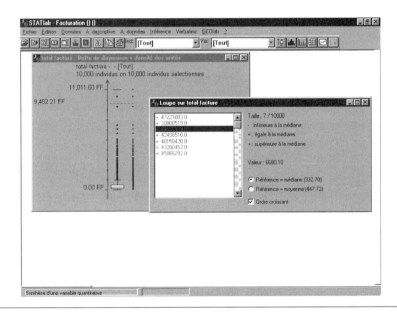

Figure A.18 Billing data by boxplot and interaction tools. (Reprinted with permission from CNET-France Telecom and SLP InfoWare Inc.)

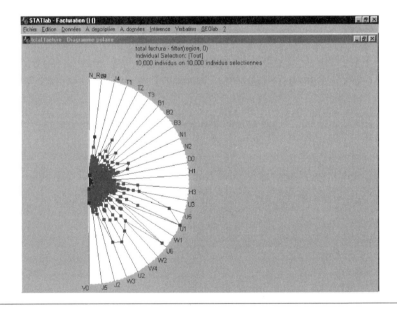

Figure A.19 Polar diagram of billing data. (Reprinted with permission from CNET-France Telecom and SLP InfoWare Inc.)

a key data report is made for the executives. The following section shows some examples of this reporting. In this case, we look at reports, created with DATAboard, that tell executives about turnover, real turnover, and turnover deviation for their areas. This turnover is given in terms of components and in regions. DATAboard also allows executives to automatically update data sets and graphics as soon as they arrive in the system. After the reports are reviewed and corrected, they can be disseminated throughout the organization via the corporate computer network.

Real-time data access and visualization

For some applications, it is necessary to have results (data visualization statistics) as soon the data are updated; every month, week, or second, depending on the type of data and usage. Many applications might be overwhelmed by this requirement because of the volume of data needed (turnover, traffic measurement, quality of service, financial data, etc.). An application based on the STATlab tools and principles was built to address these needs. The tool is called SPEEDY (see Figure A.20).

Visualization of technological watch and competitive environment data

Consider now the example of patents data considered part of the technological watch for a telecommunications operator. The patents are recorded in a database and are accessible through many different sources. The content of each patent is captured using both codification and textual descriptions. Using these as input, it is relatively easy to build analyzable data and statistics based on those databases. (Of course, it is possible to analyze detailed data from patents. We do not present them here because of confidentiality concerns. But anyone can perform analysis from the world patents database.)

Churn customer data and decision making

Because of the critical importance of churn information to telecommunications executives, SLP InfoWare built a complete system, the Customer Profile System (CPS). It permits the user to create, refine, and apply customer profiles and compare them against many different backdrops: sales, marketing, churn, quality of service, and network usage (see Figure A.21).

CPS is an enterprise-wide solution, designed for analysts, power users, managers, and field users such as sales teams, engineers, or technical support.

Figure A.20 Example of successive output from SPEEDY. (Reprinted with permission from CNET-France Telecom and SLP InfoWare Inc.)

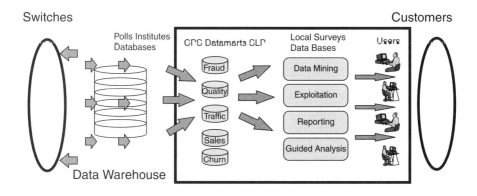

Figure A.21 SLP InfoWare Customer Profile System.

CHURN/CPS is based on SLP InfoWare's User Information System (UIS) technology that includes data mining, guided analysis, multidimensional database, dynamic object-oriented reporting, and visualization. This technology creates new data utility for businesses by utilizing innovative data mining approaches and SLP's "guided analysis," which supports the best business practices in data mining and gives users the ability to mine data without specific skills in statistics or artificial intelligence.

The Value of CHURN/CPS

Today's dynamic telecommunications environment, retaining customer loyalty is critical. In the last three years, the cost of customer acquisition has gone higher and higher due to aggressive marketing campaigns launched by small- to medium-sized competitors who are aggressively growing, and by large companies that want to both retain their current customers and suffocate smaller companies by increasing the cost associated with market entry. Three years ago the buzzword was "quality of service"; today, *churn management* has become one of the most critical business issues.

CHURN/CPS is designed to assist business units (such as marketing and sales managers, product planning, etc.) with a solution that can analyze and predict customer churn. CHURN/CPS can do this automatically, giving direct results, and helping users to understand the churn analysis and modeling process via CHURN/CPS's unique *guided analysis* module.

CHURN/CPS delivers four main types of payback:

1. Automatic modeling, or "guided analysis," to detect customers who have churned and predict customers who are highly probable to churn;
2. Direct action on high-risk customers because of CHURN/CPS's ability to identify just-about-to-churn customers;
3. Ability to launch a proactive marketing campaign based on "high-risk analysis";
4. Customers risk analysis and evaluation.

There are two CHURN/CPS possibilities: preventive CPS and proactive CPS.

Preventive CPS: analysis based on customer characteristics

CHURN/CPS is designed to help analysts or managers detect and model churning customers. From a first "test" database containing all possible information about customers as well as known churning and nonchurning customers, CHURN/CPS guides you step by step, selecting the most discriminating characteristics and creating an operational model of churning customers.

With these features, CHURN/CPS reduces the number of possible characteristics to those which are significant to the churn model, and it guarantees the accuracy of the model. Periodically, the models are updated, using the same guided analysis.

Proactive CPS: early detection of churn based on usage

CHURN/CPS has a specific guided analysis designed to detect significant changes of usage per customer. Based on pattern recognition algorithms, CHURN/CPS extracts from all usage curves (MOU per destination for instance) different profiles of curves.

The first screen, Figure A.22, shows a simple, unsegmented view of the customer population. When mixed together, we learn nothing meaningful about the churn.

The next screen, Figure A.23, shows how several different subgroups of customers correlate with each other according to intrinsic patterns.

Recognized patterns can then be projected against forecasts, allowing the company to better fine-tune its future expectations (see Figure A.24).

The models created by the guided analysis capabilities of the tool are usable at production levels for sales agents, managers, or financial agents simply

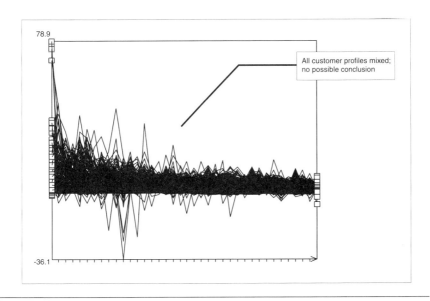

Figure A.22 Changes in usage by customers.

by including end user applets, or via intranet applications. One particular applet provides the ability to download CDR records to compute the possible churn per customer.

CHURN/CPS includes:

- Data collecting from any source (legacy systems, RDBMs, census data, etc.);
- Dynamic thresholds for an early detection module;
- Modeling and learning module with full guided analysis;
- Data mining modules;
- Reporting module;
- Web report module for intranet reporting.

Implementing CHURN/CPS

Implementation of CHURN/CPS involves four main steps:

1. Collecting and analyzing data. This first step is conducted by SLP InfoWare's database and data mining senior analysts working with the

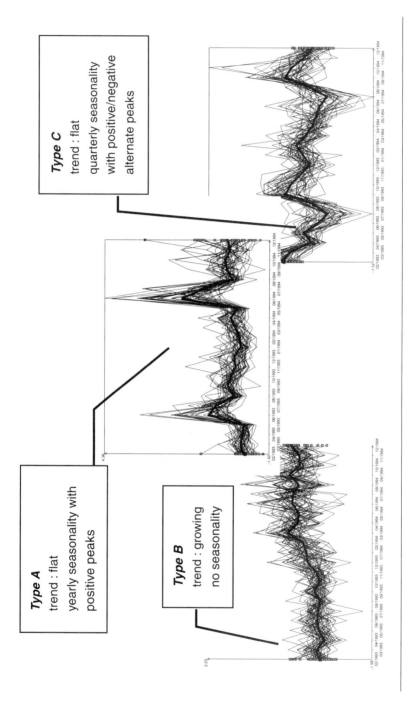

Figure A.23 Usage by subgroups of customers.

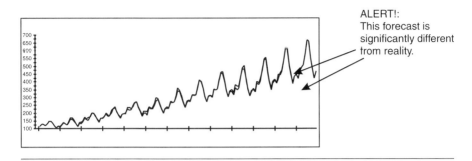

Figure A.24 A forecast that will have to be adjusted.

customers. After this step, the return on investment that CHURN/CPS can provide will be clear.

2. Establishing a proof of concept. The proof of concept is typically focused on an operational workgroup and includes data access, reporting, and guided analysis. This step will deliver complete information about users' needs and expectations.
3. Installing and configuring the complete UIS, including a multidimensional server, connection to remote sites, applet integration with other applications, and interfaces with other databases.
4. Setting up production, including training, maintenance, and technical support (software and methods).

CHURN/CPS is a near-turnkey datamart. The implementation cycle of a typical project, as outlined above, is four to eight months.

Using CHURN/CPS

Now we take a closer look at using the product. The first step (outlined previously in the section *Preventive CPS: analysis based on customer characteristics*) is to indicate which field contains the churn/no churn characteristic. Step 2 is to provide the complete list of all the variables that *could* indicate churn (Figure A.25).

The CHURN/CPS eight-step guided analysis (see the section *Proactive CPS: early detection of churn based on usage*) directs the user from rough data to model of usage, and finally, to early defection detection. Detection based on computed and stored models can be performed every day on strategic accounts

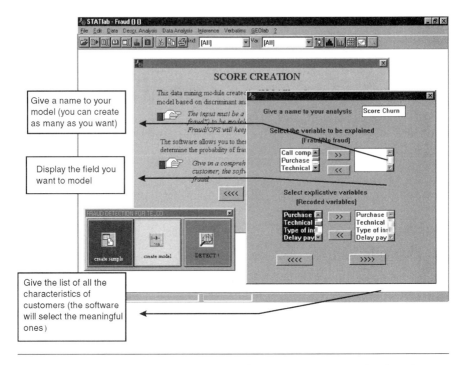

Figure A.25 Steps in implementing CHURN/CPS. The software provides the indicative variables and the score by customer. (Reprinted with permission from CNET-France Telecom and SLP InfoWare Inc.)

such as business customers, or with lower periodical usage for less intensive users.

The first screen, shown in Figure A.26, allows the user to define a churn category.

The second screen, shown in Figure A.27, presents some of the different displays and forms of analysis produced by the product for this churn grouping.

Figure A.28 shows different churn segments for the group.

The final result of the analysis provides:

- A list of customers identified as the most likely to churn (This list goes to telemarketing teams to address as the highest priority.);
- The forecast of usage per customer.

The churn model is geared toward a number of different user profiles:

Appendix A: Real world warehousing   195

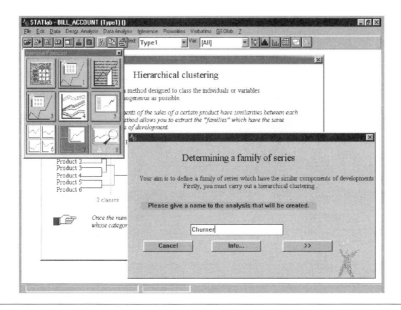

Figure A.26   Customer profiles are computed. (Reprinted with permission from CNET-France Telecom and SLP InfoWare Inc.)

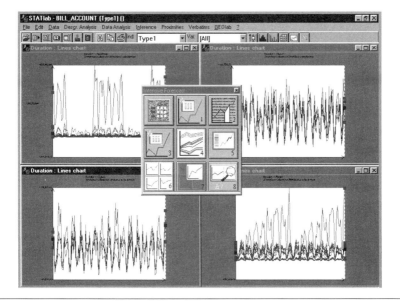

Figure A.27   Visualization of data is provided at each step. (Reprinted with permission from CNET-France Telecom and SLP InfoWare Inc.)

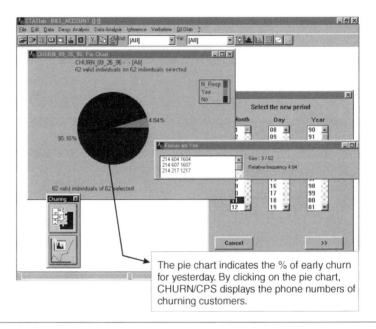

Figure A.28 Direct access to churning customer data. (Reprinted with permission from CNET-France Telecom and SLP InfoWare Inc.)

- Reports for managers (including risk evaluation reports, per type of customer, per geographic location, etc.);
- Targeting marketing campaigns for win-back mailings and preventive care of customers;
- Analysis of churn by *nonsophisticated users* such as field managers, sales managers, or regional managers.

CHURN/CPS is the essential application that will save time, target marketing actions, and preserve the base customers from competitors' initiatives.

Fraud detection

Fraud is an everyday issue for companies who deliver their products or services before payment, like power or water suppliers. Fraud is an especially critical issue for telecommunications companies because of their orientation toward service.

Usually, fraud detection systems are built directly on production sites such as switches. Due to the intensity of traffic (100,000,000 transactions or more per day), few limiting fraud rules are applied in real time to deal with the most critical scenarios. Nevertheless, there are new ways to commit fraud, along with new prevention methods and customer risk analysis, neither of which are supported by these systems.

As part of the customer profiling system, a fraud detection component was built, called Fraud/CPS. This component was designed to fill these needs, delivering three main types of payback:

- Customer risk analysis and evaluation;
- Automatic or guided analysis of fraudulent customers, detecting new types of fraudulent customers;
- Application of new models in production systems.

Fraud/CPS helps analysts or managers detect and model fraudulent customers. The steps through the guided analysis for building a Fraud/CPS model are the same as for CHURN/CPS . Fraud/CPS includes:

- Data collection from any source of data (legacy systems, RDBMs, census data, etc.);
- Rules-based fraud detection;
- Usage-based fraud detection with dynamic thresholds;
- Modeling and learning module with full guided analysis;
- Data mining modules;
- Reporting module;
- Web report module for intranet reporting.

For the four steps to implement Fraud/CPS, you can also refer back to CHURN/CPS, as they are identical. As in CHURN/CPS, the implementation cycle for Fraud/CPS is also typically four to eight months.

Figure A.29 shows the initialization screen for this application. Notice the complete *query building* functionality offered in the Expression generator window, and the *create variable* capabilities.

Based on certain rules, customers are marked in a test database as fraudulent or nonfraudulent. At this step we also utilize corporate database(s) of past customer fraud, if it exists.

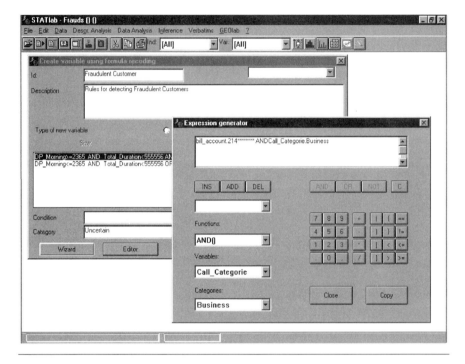

Figure A.29 Initialization of the Fraud/CPS system. (Reprinted with permission from CNET-France Telecom and SLP InfoWare Inc.)

The first step in using Fraud/CPS is to set up for score creation (Figure A.30).

The next screen (Figure A.31) shows how you can document the different variables to use and compute the scorings.

The final screen (Figure A.32) shows scores and results.

The guided analysis leads the user step by step from the initial data to the model. The first step is to indicate which field contains the fraud/no fraud characteristic, the second is to provide the complete list of all the variable that *could* indicate fraud. The software then provides a list of the most discriminating variables and the score per customer.

The features of this module include verification of the quality of the model.

The fraud model can provide information for a number of different users:

- Reports for managers;
- Online detection in call centers;

Appendix A: Real world warehousing 199

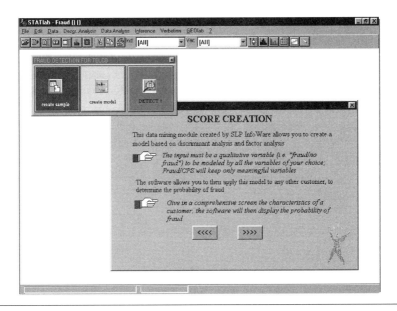

Figure A.30 First step in the fraud profile. (Reprinted with permission from CNET-France Telecom and SLP InfoWare Inc.)

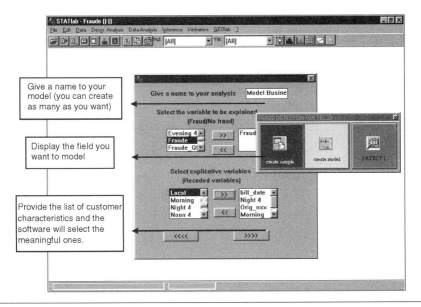

Figure A.31 Creating the model. (Reprinted with permission from CNET-France Telecom and SLP InfoWare Inc.)

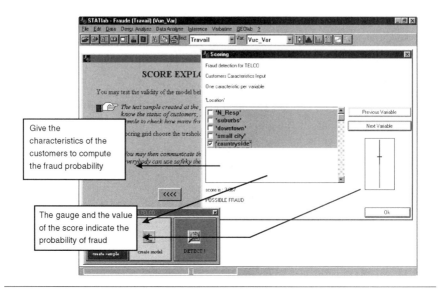

Figure A.32 Final step for Fraud/CPS. (Reprinted with permission from CNET-France Telecom and SLP InfoWare Inc.)

- List of fraudulent customers to check;
- Network-based production systems (switches).

Data availability

The data used in the business decision-making process can be alpha or numeric. Data may be obtained from questionnaires or surveys, from enterprise computer systems, or from the competitive environment. Therefore, the warehouse developer's first challenge is to figure out how to import the data regardless of the source. The STATlab solution can import data from many different types of sources (questionnaires, data sets, survey panels, samples, measurements spectrum, images, etc.). These sources can be combined for easy, transparent data analysis.

Architecture and Technical Specifications

From a technical point of view, the STATlab solution is based on recent and long-term techniques to allow maintenance and durability. The chosen standard is distributed computing environment (DCE), the standard for OSF users for client server tools. The industry-established DCE standard is widely available on the exploitation system market. DCE is of interest to end users as well as developers because it solves the communication protocol problem. Among the different services offered by DCE, CNET (SLP's solution) uses the following: (1) thread services for task parallelization and (2) remote procedure call (RPC) services for execution of procedures from a distant workstation. The following protocols are available: TCP/IP, Net BUI, Net BIOS, LOCAL, UDP, and IP/SPX. The software was developed in a multiplatform context and is available on UNIX, (HPUS, AIX) and Windows NT servers for data servers, and on Windows, OS/2, and Motif Macintosh for enduser workstations.

Due to the integration of the Q+E and ODBC technology, as well as a customized interface with Oracle, the STATlab solution can integrate and import data from most of the classic computer environments such as Oracle, Informix, Ingres, Btrieve, Clipper, Database Manager, Teradata, SQL/DS, SQL 400, Excel, DBase III and IV, ASCII, CSV, Access, Microsoft SQL Server, Progress, SQLBase, Sybase SQL Server, XBD, HP Allbase/SQL, HP Image, Image/SQL, and DB2 (MDI, DDCS/2). For more information, a special technical report is available from SLP InfoWare Inc.

Enduser tools

Following is a brief description of the tools included in the STATlab Data Warehouse Solution.

Data analysis with STATlab

The STATlab data analyzer is the heart of the STATlab environment, containing most of the data analysis and management functions, plus a data dictionary and a list of possible data access approaches. It is an intuitive, user-friendly, graphical user interface (GUI)-based, exploratory data analysis environment integrating statistics, graphics, and data management for use in market research, sales analysis, network management, customer analysis, operations management, survey analysis, and process improvement.

Its user-friendly GUI makes complicated analysis easy. STATlab's unique information object architecture makes it simple to drilldown into graphs and data. It allows for interactive point-and-click data localization and identification directly on graphs. It has visual correlation through color highlighting and brushing of data points.

STATlab is comprehensive and powerful, with more than 100 different interactive graphs and numerous statistical treatments. You can gather, classify, compare, and build models while exploring data, discover causes for variations, correlations, and dispersions in data, test differences between populations, and trace the genealogy of variables. STATlab uses function wizards to create and derive new variables, statistical assistance through online examples and advice, and automates routine tasks with STATlab's macro language.

Data analysis capabilities include univariate (simple and chronological), bivariate, trivariate, and multivariate analysis, as well as factor analyses (principal components, simple and multiple correspondence), classification analysis (hierarchical clustering, partitioning, discriminant analysis, segmentation), regression analysis (forward, backward stepwise), multidimensional scaling (MDS), and verbatim analysis.

With STATlab you can import data from most common data formats: relational databases, spreadsheets, ASCII files, and other popular statistical analysis packages. There are no limitations on the size of data sets and data elements. It has automatic creation of a data dictionary while importing data, as well as easy to edit and modify data dictionaries.

You can input, edit, and modify data directly on STATlab's spreadsheet data manager, create filters through graphical selection and formulas, and cut and paste data from other Windows applications.

Table A.1 STATlab Capabilities

Extend STATlab into a user information system with:	
DATAboard	**Dynamic object-oriented reporting**
GEOlab	Statistical mapping objects
TIMElab	Interactive forecasting
CSQlab	Statistical quality control
DATAread	Multibase querying environment
DATAman	Natural data modeling

Natural data modeling with DATAman

DATAman, a natural data model based on a client-server multidimensional-data architecture, provides data services for DATAread and STATlab. It covers the functions of data management, data server, data dictionary, data security, data modeling/editing, and data importing.

DATAman is organized for exploratory data analysis and extended online analysis processing (OLAP) architecture. It manages statistical and decisional data and integrates relational and nonrelational data sources.

With DATAman you can easily initialize and manage natural data models, organize data by time, organization and aggregation, load data through simple GUI-based import utilities, and define automated editing, validation, and computational functions.

DATAman's powerful data dictionary manages data names, aliases and codes; defines aggregation, hierarchical levels, chronological and nonchronological variables; and establishes automatic data validation criteria.

DATAman has extensive data security. It controls access through password facilities, limits access to subsets of data by user, and restricts permission for read/write access.

DATAman allows multiple mergers and concatenations of data, supports multilevel aggregation, automatically validates imported data, automatically computes new variables and aggregations, and updates computations while importing new data.

DATAman allows import of data from most common data formats, supports most UNIX environments and Windows NT, is LAN and WAN network protocol-independent, and has a client-server architecture based on a distributive computing environment (DCE).

A multibase querying environment with DATAread

DATAread is an easy-to-use, GUI-based data querying and access environment for relational databases and DATAman, addressing the functions of query environment, data entry/editing, query builder, data import/export, and query management.

DATAread creates STATlab data folders directly from popular relational database systems. The point-and-click SQL query builder makes it easy to import data subsets. It automatically creates and catalogs most frequently used SQL queries, making it easy to perform periodic routine and repetitive tasks. Cataloged SQL statements can easily be edited and customized to meet unique requirements. A user can import data from several databases in the same format through SQL queries.

You can create STATlab data folders directly from DATAman's Natural Data Model through a GUI-based, natural query builder. It has point-and-click selection of key elements of a natural query: units, variables, and time periods. You can select from a list of units, individually or by groups, taking advantage of the natural hierarchy of the database. There is also a list of options and variables within each domain or subdomain.

DATAread can catalog and save queries by name for routine and repetitive tasks (Figure A.33), edit and customize cataloged queries to meet unique requirements, close or delete any query, test a data query before execution to assure accuracy, performance, and validity, and maintain an updated list of all the data available on the server.

Figure A.33 shows DATAread query results displayed in an Excel spreadsheet. DATAread and DATAman provide a data-, network-, and plat-

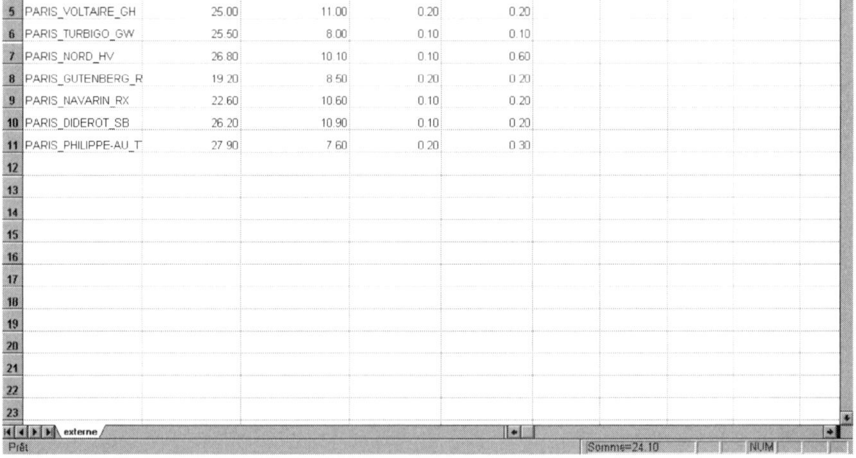

Label	SINUM_9510	SILLUPT_9510	SITFZ_9510	INCIDENT_9510
PARIS_NORD_U5	24.10	9.50	0.10	0.30
PARIS_ANJOU_BR	20.30	9.90	0.20	0.20
PARIS_MONTMARTRE_	27.70	8.00	0.10	0.20
PARIS_VOLTAIRE_GH	25.00	11.00	0.20	0.20
PARIS_TURBIGO_GW	25.50	8.00	0.10	0.10
PARIS_NORD_HV	26.80	10.10	0.10	0.60
PARIS_GUTENBERG_R	19.20	8.50	0.20	0.20
PARIS_NAVARIN_RX	22.60	10.60	0.10	0.20
PARIS_DIDEROT_SB	26.20	10.90	0.10	0.20
PARIS_PHILIPPE-AU_T	27.90	7.60	0.20	0.30

Figure A.33 DATAread query results displayed in an Excel spreadsheet. (Reprinted with permission from Microsoft Corporation.)

form-independent client-server OLAP, multiple dimension database (MDD) environment. DATAread provides security for accessing both DATAman and external relational databases. DATAman allows users to configure and modify natural data models. DATAread supports security through user name and password on the server.

You can create data entry templates for entering data directly into DATAman or STATlab, load, edit, and modify server data on the client computer and send it back to the server data-base, and import/export ASCII data files and spreadsheet data.

Statistical mapping objects with GEOlab

GEOlab is an interactive, easy-to-use, GUI-based statistical and object mapping environment integrating maps, statistics, and data management, used for territory planning, market surveys, environmental studies, sales analysis, and semiconductor wafer mapping.

GEOlab allows you to visualize the results of the STATlab statistical analysis on maps and objects; cut and paste data from Windows applications; import data from spreadsheets, relational databases, ASCII files, and other popular statistical analysis packages; and aggregate and summarize data and display them on maps and objects.

GEOlab is integrated with DATAboard's dynamic object-oriented reporting environment for complex querying, reporting, and high-quality presentations. You can cut and paste GEOlab's maps and objects into other Windows applications, point, click, and drilldown through GEOlab's maps and objects to analyze data, and visually correlate data on GEOlab's maps and objects for quick results.

GEOlab has numerous graphs and objects, and supports their surface coloring based on data and analysis results. GEOlab displays underlying data values on maps and objects, as well as through bars, surfaces, volume charts, pie charts, histograms, hemicycles, sun ray plots, and profile diagrams.

Interactive forecasting with TIMElab

TIMElab is an easy-to-use, GUI-based, interactive environment for time-series analysis, modeling and forecasting integrating statistics, modeling and data management It is used for sales forecasts, business modeling, budget projections, inventory modeling, investment analysis, and process modeling.

TIMElab has a user-friendly environment with pop-up menus to guide you through analysis. It lets you immediately see the results of interactive modeling efforts, intuitively search for the best model without advanced statistical training, point and click to project your forecast, and cut and paste your forecast results into popular Windows spreadsheets.

You can model your indicators with simple and double exponential smoothing, use Holt-Winters exponential smoothing, both multiplicative and additive, multiple regression with environmental indicators and confidence intervals. TIMElab lets you create and customize your own moving average formula, apply or remove seasonal factors to improve your forecasting model, and utilize auto-correlation functions and series transformations and the advanced statistical analysis features of STATlab.

TIMElab can import data from most common data formats: relational databases, spreadsheets, ASCII files, and other popular statistical packages. There are no limitations on the size of data sets and data elements. TIMElab allows for automatic creation of a data dictionary while importing data. It has easy to edit and modify data dictionaries. You can input, edit and modify data directly on STATlab's spreadsheet data manager, create filters through graphical selection and formulas, cut and paste data from other Windows applications, and use DATAread to query and access data from external relational databases and DATAman.

For more information on data management and forecasting capabilities request product literature from SLP InfoWare Inc.

Statistical quality control with CSQlab

CSQlab is an easy-to-use, interactive environment for implementing applications and performing statistical quality control analysis integrating statistics, charts, graphs and data management, for quality control, management, statistical quality control, statistical process control, quality control analysis, and quality control reporting.

CSQlab has a user-friendly environment for developing and analyzing charts. It allows you to point and click box plot analyses of subgroups directly on charts, easily identify out-of-control subgroups, and color highlight individuals through simple menu selections. It has pop-up menu access to moving average, warning curves and characteristics tables, and interactive drilldown and analysis of underlying data.

With CSQlab you can use Pareto charts, Ishikawa diagrams, and univariate and bivariate statistical analysis for selecting control characteristics; create and develop control charts customizing warning and control limits; use normal distribution assumption tests to develop control chart strategies; add new points on previously computed control charts; and cut and paste CSQlab objects into the most popular Windows applications.

CSQlab has numerous control charts: (X,R), (X,S), (X,S^2), (X,MR). Chart S2, CUSUM, CUSUM with V-mask, moving average, and EWMA. It has charts for attributes data: np, p, c, and u charts, as well as integrated extensive statistical analysis capabilities: univariate analysis, univariate chronological analysis, and bivariate analysis. You can use STATlab to perform advanced exploratory data analysis.

CSQlab supports import of data from most common data formats: relational databases, spreadsheets, ASCII files, and other popular statistical packages (Excel, Lotus, dBase, ASCII, SAS, Statgraphics, SPSS, Systat, Minitab, ITCF). There are no limitations on the size of data sets and data elements. CSQlab can automatically create a data dictionary while importing data, and has an easy to edit and modify data dictionary. You can input, edit, and modify data directly on STATlab's spreadsheet data manager, create filters through graphical selection and formulas, cut and paste data from other Windows applications, use DATAread to query and access data from relational databases and DATAman, and use SPEEDY from SLP InfoWare to acquire real-time data.

Dynamic object-oriented reporting with DATAboard

DATAboard is an easy-to-use, GUI-based, object-oriented reporting and presentation environment, dynamically linked to Natural Data Models and Exploratory Data Analysis, for online reporting, querying tools, online monitoring, publishing software, applications, and spreadsheets.

It is simple to create reports: just point and click with DATAboard's easy-to-use, GUI-based, object-oriented environment. DATAboard has an extensive tool palette for creating, designing, and building reports and presentations.

DATAboard makes complex reports easy to manage. It has numerous graphical objects, macro buttons, zoom, drag and drop, and color customization. You can illustrate reports with graphs, synthetic diagrams, maps, and reading guides (arrows, symbols, images) with user-friendly layout functions.

DATAboard's reporting objects may be dynamically linked to analysis and data. When data changes, DATAboard reports can be automatically updated and distributed throughout the network. It supports open data access to most common data formats: relational databases, ASCII files, spreadsheets, and other popular statistical analysis packages. Users can read reports page-by-page and print or modify them directly on the screen.

DATAboard includes all the common calculations: mathematical and time processing functions, moving average, and variable recoding. Users can implement intelligent icons that react to changes in data through easy-to-use macro facilities. DATAboard includes word processor functions, publishing and statistical packages, and a querying tool. It can be integrated with the statistical and analysis power of STATlab and the statistical object-mapping capabilities of GEOlab.

SCRIBE-Tutor and SCRIBE-Questions

SCRIBE is a computer-aided system for learning and teaching statistics. It is divided into two parts. The first, SCRIBE-Tutor, is a dedicated, statistical methods teaching system. The second, SCRIBE-Questions, provides users with the ability to interactively learn more about statistics. SCRIBE runs on any IBM or IBM-compatible PC and can be executed under MS-DOS and WINDOWS 3 or higher (with a VGA color monitor).

SCRIBE-Tutor contains 11 chapters: univariate, bivariate, n-variate analysis; factor analysis, principal components analysis, 2-D and N-D correspondence analysis; classification, proximity analysis, and data analysis strategies. It has a subject index with 300 entries and a reference bibliography. SCRIBE-Questions contains nearly 100 questions with multiple-choice responses. The questions coincide with the SCRIBE-Tutor material.

SCRIBE-Tutor is a 700-page multiaccess electronic book. The main screen contains an icon-graphic menu bar. The current screen is divided into two parts, a graphics page and a text page. The system is accessible by chapter, table of content section, or the index. SCRIBE is connectable with STATlab so readers can tie exercises to any real statistical work they are doing.

From data to decision

The basic reason for implementing the STATlab Data Warehouse solution is to view data in some context *that yields information*. With this information, end users can make more informed, more appropriate decisions that provide them with a business advantage.

The process involved in decision-making has been the same since ancient times. One has an objective (a hypothesis to check, a result to confirm, a situation to highlight, a target to reach). This objective is translated in terms of data, and the enduser explores this data with a set objective until a solution is found. It is not a numerical method for solving equations, but a knowledge process, based on analyzing data step by step.

Conclusion

S.G. Eick and D.E. Fyock write in *ATT Technical Journal* (1996), "Corporate databases have been recognized as strategic assets, and a successful corporation will make full use of its data resources to gain competitive advantage and to better manage its business. Visualization is a key technology for extracting information from data and is, therefore, becoming increasingly important in our information rich society. It complements other analytical, model based approaches and takes advantage of human pattern perception. Visualization can help users to navigate and explore the fast-growing number of data warehouses to more easily and rapidly discover the information hidden within volumes of data."

During the past several years, CNET and SLP researchers have developed an innovative technology that permits interactive analysis of corporate data sets whatever their size. They have built a suite of applications based on innovative, novel views, using a common software infrastructure called the *STATlab System*.

Beginning with the basic STATlab, they built a complete platform, called User Information Systems (UIS). Then, they built a range of specific computer applications such as fraud detection, churning, and customer profiling. CNET and SLP plan to continue researching and exploring this type of information technology for France Telecom and any other telecommunications operator and similar types of companies, such as banks, insurance companies, and retailers.

Contributors

Michel Jambu is the Statistical Information System supervisor of the Center of Research for France Telecom (CNET). He is also a professor and thesis supervisor at the University of Paris IX Dauphine. He can be reached at

> France Telecom-CNET
> 38-40 rue du General Leclerc
> Suite 409B
> 92131 Issy les Moulineaux
> France
> Phone: (33) (1) 45 29 50 00, 46 60 52 31
> Fax: (33) (1) 45 29 65 57, 46 60 52 31

Jean Schmitt can be reached at

SLP InfoWare Inc.
200 West Madison Street
Suite 2150
Chicago, IL 60606
U.S.A.
Phone: (312) 407 6582
Fax: (312) 407 6581

Appendix B

The business case for business intelligence

by Elvin J. Monteleone, Vice President, Professional Services Alliances
Seagate Software

GENUINE BUSINESS NEEDS are shaping the overall business intelligence marketplace, leading to the present demand for a new kind of software solution for the entire enterprise. For telecommunications companies, the capabilities of these new systems are "heaven sent," offering the answer to a most important question: once you have created your clean, well-lighted data warehouse, how do you put the data to use? The simple queries that can be handled directly by a relational system are necessary and valuable, but users' needs are much broader. Business intelligence supplies the extra dimensions that can make proprietary data a competitive weapon for the telecommunications industry.

Enterprise business intelligence in particular is appropriate for data-intensive, competitive industries such as telecommunications. Enterprise business intelligence necessitates an all-encompassing view of an organization's business needs in order to leverage corporate data for maximum return. The enterprise business intelligence approach starts with the elements that drive the business itself and extends to the way in which the business intelligence system is developed and deployed, including leveraging the existing information technology (IT) infrastructure. A good enterprise business intelligence system, such as can be implemented with Seagate Holos software from Seagate Software, systematically delivers the right information to the right people across the organization at the right time. In the course of doing this, it matches business processes (either existing or re-engineered) and produces competitive advantage for the organization.

Today, we see the beginning of a convergence between conventional line-of-business data processing and the specialized processing previously used to inform and enhance management decisions. At the business level, it makes sense not just to inform, but to empower the manager—to feed back the enhanced decisions of the individual into the operational systems of the business. This is the future direction for business intelligence, which will both open a new and very large marketplace for enterprise business intelligence systems and enable user organizations to improve profitability, sometimes very substantially.

B.1 Our credentials: background of Holistic Systems/Seagate Holos

Holistic Systems, now part of Seagate Software, has been in the business intelligence marketplace since 1988, and we have a number of firsts to our credit. We believe we are well placed to make useful—and credible—observations about the business intelligence marketplace and relevant technology:

- Seagate Holos software, introduced in 1988, was the first product to combine so-called executive information system (EIS) and online analytical processing (OLAP) technology in a single integrated offering. Some EIS purists at the time criticized the approach as being "improper" EIS. Indeed, a number of vendors initially suggested that EIS and DSS (OLAP) were so radically different that they should never be considered in the same breath. Events have proved them wrong.
- Seagate Holos was the first business intelligence product to use the "client-server" model. Indeed, the term didn't exist at the time, so it was

called "cooperative coprocessing," which, in fact, later became an accepted term for one form of client-server architecture.

- Seagate Holos was the first business intelligence product designed to integrate its OLAP engine directly with standard relational databases. Seagate Holos does not require a proprietary OLAP database—it's "open." When Seagate Holos was first introduced, this ability to work with any relational database wasn't described as relational OLAP (ROLAP) because the term hadn't been invented, but it was clearly delivering both trend-setting technology and real customer benefits as early as 1988. Seagate Holos, in fact, goes beyond ROLAP and could more properly be termed "HOLAP," or hybrid OLAP, for its optimal combination of multidimensional and relational capabilities.

- Although its original architecture was designed in 1987, Seagate Holos remains at the forefront of business intelligence technology through a continuing aggressive development program. Seagate Holos 5, released in mid-1996, is essentially a new product, with over 70% of the system having been rewritten in the previous two years. This was achieved while maintaining complete upward compatibility, so that Seagate Holos applications written in 1988 continue to run on current versions with minimal changes—an extraordinary commitment to its customers.

- This commitment also assures Seagate Holos users that the systems they bring online today will work with the business intelligence enhancements of tomorrow. With the approaching convergence of telecommunications and other media, the future holds synergies and challenges undreamed of in the present—new applications, new demands for data, new markets, customers, and competitors. (It also holds vast challenges for information systems (IS) who will have to provide information workers with consolidated data that may be the result of strategic alliances, mergers, acquisitions, and so on.) By building the most flexible, extensible business intelligence systems possible today, using Seagate Holos, telecommunications companies gain a competitive edge—confidence that they will be able to rise to tomorrow's challenges and solve problems they can't yet even imagine.

B.2 Technology alone is the wrong emphasis

Historically in the business intelligence market, too much prominence was given to technology over business requirements. For example, many software vendors, including Holistic Systems, benefited from the hyperbole surrounding the early days of OLAP and were naturally pleased to have the recognition of the benefits of multidimensional data handling brought to the fore. However, this attention in many cases was derived more from the opportune marketing of the technology than from a genuine understanding of its usefulness. The truth is that OLAP isn't a single panacea, but rather one valuable element among the many needed to deliver certain classes of application.

To realize the full value of OLAP and related technologies, it's necessary to take a step back from the technology to look at the underlying business issues. First, take the EIS market. With hindsight, one can see that that the original idea of providing a highly graphical reporting system, aimed at a small number of board-level executives, delivering very high-level historical data, did little or nothing to make the organizations that used them more competitive, and as a result these systems were of limited value.

This does not mean that the technology used to deliver EIS was bad. It was just that as a result of looking at it from the wrong direction, it was most often applied in a manner that offered little real benefit to its users.

If we take another step back and look at the business intelligence systems marketplace as a whole, we have the opportunity to ask, "How can we use this technology to make a real difference?" The emphasis here is crucial. We aren't looking at how to force-fit a particular technology into an organization, but rather trying to find a way to deliver information more effectively to real advantage. Only when this is decided should we survey the technologies available and select the ones that fit with the answer.

B.3 Focusing on the business is the right emphasis

Most people who have been involved with information technology for some time have become altogether too familiar with the line that the benefits of a system are either "intangible" or worse, "impossible to measure." This is usually merely a dead giveaway that it is a technology-driven system. If it isn't possible to determine beforehand where the "bottom line" benefits of a system implementation lie, then we should ask whether it should be started at all.

While it may be difficult to quantify the benefits of systems that keep the executive better informed, streamline the reporting cycle, and facilitate communication between management, it should not be difficult to assign numbers to systems that deliver information allowing the following:

- Call center managers to reduce waste;
- Purchasing managers to get better prices from their suppliers;
- Direct marketers to better target their marketing;
- Traders to better understand the structure of their deals;
- Product managers to make better pricing decisions.

The closer a system is to the operational edge of the business, the easier it is to draw a direct line from the system to the real, "bottom line" benefit. Although the nature of these systems will vary among organizations and among industries, the trend now developing shows that enterprise business intelligence systems will deliver real benefit and competitive advantage over the next few years.

B.4 What distinguishes enterprise business intelligence?

If these enterprise business intelligence systems vary so much between organizations, what can we say about them in general? They have a number of distinguishing features:

1. They must have measurable benefits in saving or (preferably) making money.
2. They are most effectively deployed close to the operational managers of an organization.
3. They help those managers tackle the high data volumes that are common today, especially in telecommunications environments, by providing targeted information delivery rather than relying on generic query tools.
4. They are often built within a "data warehousing" environment.
5. They go beyond historical reporting to provide information that either genuinely helps the enterprise-wide operational management of the organization today or guides it to the future.
6. They require a new partnership between technologists and end users.
7. They quickly become essential to the broad operations of the business; that is, they are "mission critical," not just "nice to have."

We can look at these attributes in a little more detail as well.

They must have measurable benefits in saving or (preferably) making money. It must be clear where the benefits of these systems will lie. It is no good waving your hands and saying "all the benefits are intangible," but it is often unrealistic to expect an exact calculation of financial saving before starting. However, this does not mean that there should be no attempt to link the information to real financial benefits. It is easy to see how the following statements could be turned into concrete and realistic cases for business benefit:

"...this system will allow our buyers to negotiate more effectively with our suppliers, and each percentage point we reduce our costs by will increase our profit by X."

"...this system will allow us to better target our direct marketing, and if we get an increased closing rate of 10% then we will increase profit by Y."

They are most effectively deployed close to the operational managers of an organization. It might seem superfluous to say this, but enterprise business intelligence systems belong on the desks of operational managers across the enterprise. It is only by delivering information directly to the people who can make a real difference to the business that these systems can realize their full benefit. However, there may be hundreds or even thousands of potential users in many geographical locations within a large organization, and this must be taken into account when planning.

They are most efficient when they provide targeted information. A disadvantage of delivering information directly to the operational managers is that if we get it wrong, we will just provide another way to consume their time to little or no advantage. This is particularly true with regard to telecommunications companies' ever-increasing volumes of data being collected—"drowned in data but starved of information" may be a cliché, but it is a regrettably accurate description for many organizations.

To be effective, business intelligence systems must help their users tame the data they need to do their job. This can only be done through a well-analyzed and designed interface. It can be temptingly easy to gather together a large set of data and let the users at it with a generic query tool, but this can never be as efficient as producing a system that mimics the user's business processes and therefore "speaks the users' language." (Seagate Holos has sometimes been described as being an extension of Excel or other spreadsheets—leveraging end-users' skills with a familiar interface.) The benefits in the savings of end-user time and in the quality of the information delivered far outweigh the up-front investment.

They are often built on the data within a "data warehousing" environment. The current popularity of data warehousing in the telecommunications industry shows that companies are eager to make available, and thus benefit from, the large volumes of low-level, detailed data that usually exist. However, if this data is to be delivered to operational managers, then it must be summarized, analyzed, and presented in a form that can be assimilated by them as part of their daily routine.

In such an environment, it is extremely important to keep in mind the basic idea that we are trying to add value to the business through information, not just deliver data to it. As mentioned above, this requires an understanding of exactly which data each group of end users needs access to and how it can be best delivered to them. Initially, most end users will say that they require *all* views of *all* possible data. However, this is rarely true, if ever, and can lead to systems that never allow the users to see the forest for the trees.

They go beyond historical reporting to provide information that genuinely helps the operational management of the organization today and/or guides it to the future. There is undoubted value in having ready access to transactional information or other data representing what has already happened. But to make money from that information, it is necessary to use it to make decisions that will affect future profitability, reduce costs, or produce some other form of real payback.

Sometimes this can be in the form of up-to-the-minute reporting that really affects the decisions being made in a company. Increasingly, however, it also requires the ability to manipulate data to answer questions about what might happen in the future, from relatively simple what-if-style analysis to sophisticated forecasting techniques. Again, though, it's important to shield end users from the technological details. They should simply be provided with data manipulation functionality that appears natural to them because it matches the way they do their jobs.

They require a new partnership between technologists and end users. Much of the preceding discussion is based on the idea of end users being provided with systems that unobtrusively provide them with the data and analysis they require to do their jobs. This means that someone must deliver these systems to them, despite the recent vogue for "down-sizing" IT departments and the craving for "end-user development." We see the most successful organizations using small teams of hybrid "analyst-developers" to deliver appropriate functionality to end users quickly and effectively. These teams must have a good grounding in the technology and what it can deliver, but they must also be able to help the users understand what is possible and how it can be best used to serve them. Most importantly, they must work in partnership with the users to deliver benefit to their company as a whole.

They quickly become essential to the operations of the business, or close to it; that is, they are "mission critical," not just "nice to have." If all the preceding characteristics are present in a given system, then it's almost certain that removing the system would have a detrimental and broadly noticeable effect on the operations of the organization it serves. Though "nice to have" systems can be of benefit, there is much greater scope for the implementation of "mission critical" business intelligence systems that offer a far larger return on investment.

B.5 Talking about the technology

So far, we have established that there are some fundamental common characteristics of enterprise business intelligence systems, and that these derive from business requirements. But to achieve these characteristics, there are some technical challenges that must be met by the systems and software that are intended to deliver enterprise business intelligence.

B.5.1 Flexibility

As this type of system is best delivered "in the user's own language," flexibility is a key requirement. Although it is often tempting to pick a solution that appears to require less development (or even seems amenable to end-user development), this is almost always a mistake. The compromises required to fit the problem into the technology are too great.

B.5.2 Many users

The operational managers may number in the hundreds or thousands. It must be possible to deliver information to all of these in a controlled and robust manner.

B.5.3 OLAP and relational

Although they don't talk about it in these terms, users work best with data presented multidimensionally, and this is the reason that OLAP technology plays such a large part in these systems. Unfortunately, the vast majority of corporate data is stored in relational databases. This means that the chosen technology must be able to offer the speed and ease of use of OLAP truly integrated with relational storage and all its advantages.

B.5.4 Development tools

There is a naïve belief in the market today that new technology will soon allow the delivery of business intelligence systems without development. But it is only by developing well-engineered systems for end users across the organization that the full benefit of enterprise business intelligence will be realized. Although it is secondary to the ability to deliver effective systems to the operational managers, a robust development environment is also important. As we know from other areas of life, one size/shape/design not only doesn't fit all, it may fit no one at all.

B.6 Individual technology components of enterprise business intelligence systems

Enterprise business intelligence systems, in addition to utilizing a holistic approach to serving the business' needs, typically include many of the technology components listed in an earlier section of this book:

- *Agents*—Software constructs that allow users to specify the kind of information they want to access, and empower the computer system itself to go and find it, analyze it, and report back about what is found. Agent technology is incorporated into Seagate Holos software.

- *Query and reporting tools*—These products represent the latest generation of ad hoc access and reporting tools. These tools have been evolving greatly since the early days of Focus, QMF, and Oracle*Forms. Today's query and reporting tools make it possible for people to ask for business information in business terms with a minimum of SQL and other programming language requirements. In addition to the reporting capabilities built into Seagate Holos, Seagate Software offers the very popular Seagate Crystal Reports and Seagate Crystal Info desktop tools. (These products together, in fact, allow a wide range of business intelligence activity, starting at the desktop and extending to the overall enterprise, from simple end-user access and reporting, to complex queries, to sophisticated analytic manipulation of data and reports.)

- *Statistical analysis tools*—These products represent the latest generation of the traditional mainline statistical report tools like SPSS and SAS.

Seagate Holos has particularly robust statistical analysis and forecasting capabilities, incorporated with its easy-to-use report generator.

- *Data discovery* (known as data mining by Seagate Software)—Into this category we place all products that apply any kind of statistical, superstatistical, or artificial intelligence to the process of interpreting large amounts of data. Techniques like CART and CHAID, constructs like neural networks or decision trees, and a wide assortment of specialized tools make prediction, estimation, and forecasting a scientific as opposed to artistic process. Seagate Holos provides a broad and strong range of these capabilities.

- *OLAP*—Online analytical processing tools, usually incorporating some kind of multidimensional capability, provide business analysts with the ability to "surf" through their stores of data in search of new insights or conditions. A key feature of Seagate Holos, and perhaps one of the primary factors influencing the explosion of data warehousing today, OLAP technology is discussed in detail further on.

- *Visualization*—Products that make data easy to understand and interpret by displaying it on maps, multidimensional graphs, or other innovative techniques. These tools are just beginning to emerge into wide acceptance and usefulness.

- *Web-based discovery*—The burgeoning use of the World Wide Web has resulted in an avalanche of products to input, access, and analyze data over the Internet and corporate intranets.

B.7 A closer look at OLAP

In 1993, a "white paper" was produced by Codd and Date Inc. entitled "Providing OLAP (On-line Analytical Processing) to User-Analysts: An IT Mandate," by E. F. Codd, S. B. Codd, and C. T. Salley. Codd and Date's primary claim to fame was the work carried out defining the compliance rules for truly relational databases. The OLAP document again is defining compliance rules, but in this case the rules cover the area of "multidimensionality," a key component of modern business intelligence solutions. The document asserts that while most commercial offerings of relational databases "fall short of the relational model" as specified by Codd and others, they do provide powerful yet simple solutions for a wide variety of applications.

The document goes on to state that the commercial DBMS products do have limitations with respect to providing adequate functionality to support the different "views" on data. Further, simple front-end query and report writers and spreadsheets are extremely limited in the ways in which data that has been retrieved from the RDBMS can be aggregated, summarized, consolidated, viewed, and analyzed. The authors go so far as to say, "Most notably lacking has been the ability to consolidate, view, and analyze data according to multiple dimensions in the way that make sense to one or more specific enterprise analysts at any given point in time."

Hence arose the need for an OLAP engine that sits between the database and the end user. As noted previously, users instinctively tend to think multidimensionally, so that multidimensionality is a critical characteristic for these engines.

The Codd and Date white paper proposed 12 rules for optimally functional OLAP products, all of which are supported by Seagate Holos:

1. Multidimensional conceptual view;
2. Transparency;
3. Accessibility;
4. Consistent reporting performance;
5. Client-server architecture;
6. Generic dimensionality;
7. Dynamic sparse matrix handling;
8. Multiuser support;
9. Unrestricted cross-dimensional operations;
10. Intuitive data manipulation;
11. Flexible reporting;
12. Unlimited dimensions and aggregation levels.

Holistic Systems recognized in 1987 the necessity of a multidimensional capability for business intelligence systems (EIS and DSS, as they were then called) and the shortcomings of relational databases in this arena. Seagate Holos was designed deliberately to provide such capability, while still maximizing the benefits of the relational approach. Most importantly, it was recognized that this could and should be done without the introduction of a proprietary multidimensional database, which was, and is, considered an unnecessary and retrograde step.

It should also be recognized that an OLAP engine alone is not sufficient to satisfy the requirements of comprehensive business intelligence solutions. Sophisticated OLAP tools such as Seagate Holos contain automatic rule ordering and detection, and solution of simultaneous equations without user intervention. These are not simply "nice to have"—they are essential in any dynamic, forward-looking information system.

The ability only to slice, dice, and view historical data is clearly inadequate. These facilities, properly integrated within the OLAP environment, are even more essential given the trend towards enterprise-wide information systems with flattened management structures, where each manager takes responsibility for his or her own analytical work. The implicit suggestion within Codd and Date's OLAP document that this capability is not only separate but can be adequately carried out by spreadsheets is erroneous. Most spreadsheet-based systems tend not only to be inadequate to cope with these needs but continue to bring with them their deficiencies, even when attached to an OLAP engine.

B.8 Open OLAP—a better approach

Implicit in the Codd document is the view that a proprietary multidimensional database—often today called MOLAP, for multidimensional OLAP—is a prerequisite for OLAP compliance. Now that the clear benefits of open systems and standard relational databases have been established, this is a retrograde and unnecessary proposition, particularly when considered in the context of modern business intelligence strategies. A nonproprietary, open approach, such as provided with Seagate Holos, offers much more benefit especially when building for the future.

The Seagate Holos approach ensures that the information system remains in step with underlying feeder systems and that any changes in those systems are automatically reflected. This removes the need to separately maintain the multidimensional database and rebuild it every time the underlying data structures change, a process that can be complex and consume enormous amounts of resources. Seagate Holos applications, closely linked to underlying relational databases, have been in successful use within its worldwide customer base for a number of years, as shown in Figure B.1.

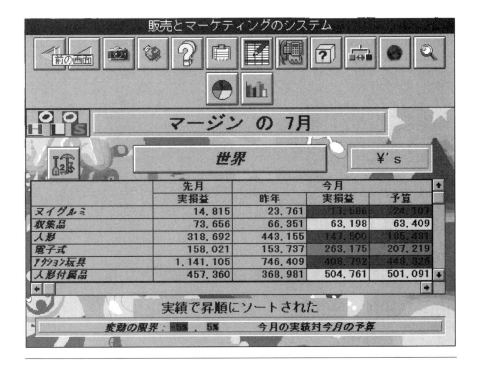

Figure B.1 Seagate Holos business intelligence software from Seagate Software is available in a Japanese language version, as well as in English, French, and German to suit the needs of large, multinational corporations. (Reprinted with permission from Seagate Software.)

Further, true databases suitable for storing large volumes of important management information require a robustness not typically available in proprietary multidimensional OLAP databases. Requirements difficult for these products to meet include proper multiuser access and update controls, recovery facilities, and audit trails. As most relational databases do provide such facilities, the Seagate Holos approach ensures that large-scale solutions can readily be accommodated, and in an environment where the appropriate levels of security and robustness are assured. To put it informally, Seagate Holos believes in "keeping the data where data belongs—in the relational database."

Large-scale information systems rely on the integrity and timeliness of the information. The major cost of most systems of this type is in the area of data management and system maintenance. The introduction of a separate multidimensional database unnecessarily complicates these key elements, causing substantially increased maintenance costs and the continued risk of the information being out of date and hence unusable. Seagate Holos' hybrid approach, its combination of multidimensional and relational storage and manipulation of data, is a solution to build on for the future.

B.9 Seagate Holos' architecture

Seagate Holos is designed to allow rapid development of large-scale, easy-to-use enterprise business intelligence applications. This class of application combines traditional OLAP multidimensional data access and reporting with forward-looking data analysis, data mining, and forecasting techniques. Coupled with an extremely flexible implementation language, this allows the creation of applications that meet the exact needs of the organization and can become key parts of its operational, profit-making information systems.

Since its inception, Seagate Holos has employed a server-centric, thin client architecture. The server machine provides the main data access, modeling, forecasting, multidimensional manipulation, and reporting capability, and acts as an "engine" for the Seagate Holos client, basically a presentation layer that runs on Microsoft Windows or Apple Macintosh computers. Seagate Holos supports mixed client populations as well, which can be important to an enterprise where the marketing department uses Macintoshes and other areas of the company use IBM-compatibles.

Seagate Holos offers a set of tools designed both to allow the development of powerful, maintainable, and easy-to-use applications and to satisfy the ad hoc needs of individual users. Server and client machines communicate through a wide range of industry-standard communications systems to achieve cooperative coprocessing, the most effective form of client-server computing in this application area. A single Seagate Holos client can connect to multiple servers concurrently.

The client-server architecture of Seagate Holos provides a number of key advantages, including the centralization of data storage (transparent access across client and server filestores), fault tolerance, easy maintainability and distribution of applications, and rapid accommodation of organizational changes. This architecture makes practicable the deployment of the large-scale enterprise business intelligence applications for which Seagate Holos is designed.

B.10 Thin client with agents, neural nets, and more

Seagate Holos is an integrated client-server tool for the development, delivery, and operation of enterprise business intelligence and OLAP applications. It unifies diverse computing facilities and data sources to provide an integrated information environment for large organizations. Seagate Holos was originally developed by Holistic Systems, an international company founded in London in 1987. In June 1996, Seagate Technology acquired Holistic Systems, which is now part of Seagate Software.

Corporate data has traditionally been stored in a variety of flat files and two-dimensional databases. As organizations move away from mainframe-based MIS (management of information systems) architectures and toward client-server environments, information consumers at all levels of the organization require access to corporate data to support decisions in a rapidly changing business environment.

To satisfy the decision support requirements of business users, data must be available in a multidimensional business view. Relational databases and data warehouses facilitate this process by consolidating, summarizing, and staging disparate data in a conveniently accessible repository. But users need powerful access, analysis, and reporting tools to obtain value from the data warehouse. Seagate Holos provides flexible data access coupled with powerful multidimensional analysis and reporting, which places actionable business information in a context that makes sense to users.

Because Seagate Holos transforms large volumes of data from disparate sources into a consistent set of meaningful information, managers and decision makers throughout an organization are able to examine historical results as well as perform complex future-oriented analytical tasks, including forecasting, "what if" simulations, trend analysis, and goal seeking (see Figure B.2).

Because the most accurate, most recent information exists in the operational systems of an organization, it makes sense for management systems to access operational information efficiently. In contrast to other OLAP products, Seagate Holos integrates tightly with the underlying relational data warehouse and legacy systems. This allows users to access, perform multidimensional analysis, and report on information from underlying data sources without the need for an additional proprietary database. Holistic Systems pioneered this approach in 1988, and today Seagate Holos, as the only HOLAP product on the market, offers the most mature, comprehensive, open-systems OLAP solu-

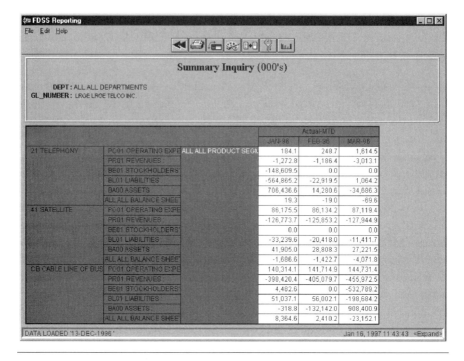

Figure B.2 A "snapshot" summary of month-to-date revenue and expense information by business category: Seagate Holos business intelligence software's tight coupling with the RDBMS allows this telecommunications company to monitor ongoing activity. (Reprinted with permission from Seagate Software.)

tion for enterprise-wide management and executive information collection and delivery (see Figure B.3).

Seagate Holos is based on a server-centric, thin-client, client-server architecture, suitable for two- or three-tier applications. In this approach, all of the analytic processing is performed on scaleable servers that are readily accessible to users throughout the organization, while the users interact with the system through modestly configured local client computers. This enables Seagate Holos to address very large-scale data volumes, to share the applications and information across the enterprise without duplication, and to avoid hardware and network capacity constraints. More importantly, users gain access to a powerful corporate-wide information processing system without needing to know where particular processing is being performed.

The combination of a purpose-built fourth-generation language and state-of-the-art graphical interfaces allows Seagate Holos to provide the best possible

Appendix B: The business case for business intelligence 227

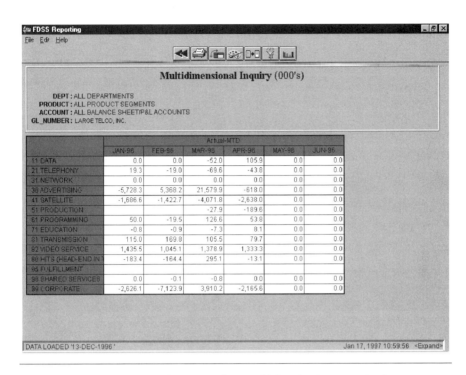

Figure B.3 Multidimensional inquiry in Seagate Holos: A telecommunications company's monthly financial information, categorized by product and department codes in the general ledger system, can be manipulated and analyzed to support informed decision making. (Reprinted with permission from Seagate Software.)

admixture of power with ease of use. A rich development environment can be used to create exactly the application needed to meet business requirements, whether it is a top-level summary for senior management or a technical working environment for the financial analyst.

The most recent release of Seagate Holos—Version 5.0—offers two flexible approaches to data mining. Pattern matching, implemented through a simple dialog interface, is a set of powerful tools to recognize patterns within data sets. Patterns include outlier, gradient, spike, step, periodicity, noise, and trend, all with configurable parameters. These pattern-matching techniques are also available as agents.

Neural nets capability is implemented as a block in the Seagate Holos 4th Generation Language (4GL), creating a radial basis function type neural net. The net is suitable for prediction and classification, and in general uncovers nonobvious relationships between input factors and some set of outcomes. The data analyzed may be in a Seagate Holos multidimensional format or may exist in relational database tables (see Figure B.4).

Seagate Holos is available on a wide variety of servers, including NCR WorldMark, Digital VAX VMS/Alpha VMS, ICL DRS6000, IBM RS/6000, HP 9000 (800 series), Sequent Symmetry, Sun Solaris, and Windows NT (Intel and Alpha). It can be accessed through MS Windows, MS Windows NT, or Apple Macintosh client desktop environments. Seagate Holos integrates large stores of information through full dynamic SQL links to most popular relational databases, including Ingres, Informix, Oracle, Rdb, and Sybase, and through open database connectivity (ODBC) in the NT environment.

In addition, data of all kinds, in any computing environment, can be quickly and easily accessed, assembled, and analyzed by Seagate Holos through its support for all popular data access and communications methods. Because

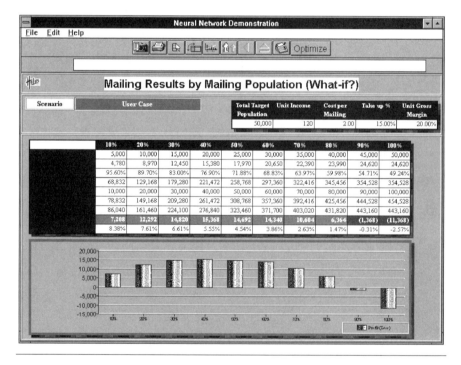

Figure B.4 Data mining capability in Seagate Holos business intelligence software from Seagate Software, either via neural networks (an example is shown above) or pattern matching, offers the automated discovery of previously unknown information within large quantities of data. (Reprinted with permission from Seagate Software.)

Seagate Holos supports many open systems' interoperability standards, leveraging existing investments in client-server architecture and relational database technology, it is an ideal component of a corporate data warehouse solution.

Seagate Software provides a full range of support services, including training, documentation, telephone and on-site support, application development, and consulting services. The company has formed strong partnerships with consulting organizations and hardware and software vendors to ensure well-supported installations.

Seagate Holos is used to provide a wide range of application solutions, including budgeting and forecasting, financial consolidations, strategic planning, product line profitability, activity-based cost analysis, financial modeling and analysis, customer service, human resources management, and marketing promotions. Customers such as Advanta, Ameritech, AT&T, British Telecom, Cisco Systems, CP Rail, Dell Computers, Walt Disney, Eurotunnel, Federal Express, Fidelity Investments, Hewlett-Packard, KPMG Peat Marwick, MBNA, Nike, Nissan, Northern Telecom, Pacific Bell, Southwestern Bell, Subaru, Transamerica Financial Services, and the University of Massachusetts have successfully applied Seagate Holos to business solutions in diverse industries, including electronics, finance, health care, oil and gas, publishing, retail, telecommunications, transportation, and utilities.

B.11 Case studies

Now let's look at how telecommunications companies such as British Telecom and Ameritech are using Seagate Holos to perform slice-and-dice interactive querying, generate just-in-time reports—saving more than a few trees in the process—and deliver enterprise business intelligence that can be used both for tactical and strategic advantage.

Basically, the dynamic nature of Seagate Holos reports, coupled with the multidimensionality of the business data being analyzed, makes it easy for users to customize "starter" reports created by MIS.

B.12 BT uses Seagate Holos for project tracking

Project tracking? Surely you jest. Isn't that a bit of, well, overkill, implementing a data warehouse and sophisticated decision support software like Seagate Holos just for project management? Not if you're British Telecom (BT) with some 10,000 ongoing projects—all of which have budgets, timelines, people, documents, and sometimes legal and other regulatory issues associated with them. Projects in the BT world are basically any organized effort that has a budget attached to it, not just IS projects.

BT began rolling out Interactive and Reporting Information System (IRIS) during Summer 1996. IRIS, which runs on a Sun SPARC platform, consists of an Oracle7 database that contains the project information along with Seagate Holos decision support software. As of January 1997, the database contained some 12 GB (gigabytes) of data and was growing at a rate of 1.4 GB per month.

The goal of the system was basically to save money by increasing efficiency. "As European markets deregulate, we are going to have to be smarter and faster," says Mike Gibson, Data Warehouse Implementation Manager at BT. "That means having up-to-the-minute information and being confident about its integrity and consistency, so that accurate decisions can be made." Gibson estimates that BT is tracking 25% to 30% more revenue-earning projects than it would have been able to before the new system was implemented—without having had to hire more employees.

B.12.1 Project rationale

BT's radical overhaul of its information systems began three years ago when the company took a long hard look at procedures in its development and procurement (D&P) arm. "There were about 2,000 development projects underway," says Gibson. "They had a lot of overlap, and although the projects often had common business themes, no one was actually managing them in a common business direction. Project managers found it difficult to work out soon enough whether they were overspending. In short, BT wasn't getting all the business benefit it should from its development efforts."

On the basis of the architecture and systems it already had, BT introduced the concept of program management, which allowed projects with a common business theme to be grouped together. But to make the most of this restructuring, the system needed to include more sophisticated analysis as well as reporting capabilities.

The project team rejected the option of bolting more bits onto their nine-year-old system, which used a reporting system based on Information Builders Focus product. This delivered paper-based and screen reports, but offered few drill-down, interactive capabilities. Also, customizing Focus reports wasn't easy for end users, since it required programming in Focus' 4GL. The BT project team went in search of a new reporting tool. "It had to provide pre-canned reporting, data analysis, modeling and trending," Gibson explains. "We were looking for a client-server application that would run on both Macs and PCs, with all the number crunching done at the server end. I wanted to avoid the need for an intermediate layer such as Oracle SQL*Net (middleware) because of the integration costs."

After evaluating almost 40 products, the project team chose Seagate Holos, a multidimensional reporting tool developed by Seagate Software and, over a period of 18 months, built the Flexible Interactive Reporting Environment (FIRE). FIRE is a story in itself. The FIRE team had decided they wanted to deliver a proof of concept preproduction system in six weeks. That would build their internal customers' faith in their ability to deliver OLAP/data warehousing solutions to business needs. Team members were given a one-week Seagate Holos training course, and Seagate Holos consultants were used to design the data structures within the system. Coding the preproduction system only took two and a half weeks, with the remainder of the time devoted to refining end-user requirements. The team had to take a customer-focused view, understand the users' roles, and replicate the kind of behavior they needed to move to.

An important part of the initial system was building extensive help files—clearly project managers couldn't be expected to attend one-week Seagate Holos training courses—and the team spent many days defining the various terms used by the system. Senior management was kept apprised of progress in an effort to keep them involved and enthusiastic. One team member recalls that "When the first demonstration was given, the audience (which included several of those senior managers) were quite skeptical at first, but as they warmed to the presentation, they got up out of their seats and gathered around the large monitor." Their enthusiasm was harnessed as renewed management support. Many volunteered to be alpha testers, since the Seagate Holos front end offered them so much more flexibility than the traditional executive briefing books and green bar reports they were used to. The team's main problem was that they wanted the system there and then, rather than waiting for the full rollout.

The original FIRE production system was implemented on a multiprocessor SPARCserver 1000 and supported some 900 users. The Seagate Holos data was refreshed nightly. The Oracle data warehouse allowed users to combine financial information—especially costs—with project effort measured in person days, highlight reports, and project milestones. The system is used to transform data in the warehouse into business intelligence, covering such areas as project status, customer satisfaction, performance measurement, plans, budgets, and forecasting. It also provides full modeling facilities.

Project managers could also enter project reports directly into the system, including both textual and numerical data. The Seagate Holos Editor was used originally, but a link to a word processor has since been introduced. Seagate Holos automatically launches Microsoft Word and saves the text in the Oracle database when the report has been completed. The system then shows the status of each project. Users can drill down to increasing levels of detail through to individual transactions. One feature users found particularly useful was the ability to make forecasts by dragging points on a graph, while Seagate Holos automatically changes the underlying numbers.

The system architecture was open and extensible, with any authorized user having read-only access to all data within the system. However, data owners could write to their part of the system. Since the original launch, the system has evolved to integrate messaging, include daily tips, and so on.

"Seagate Holos met all our requirements and gave us headroom with respect to capacity and functionality," says Gibson. "Not many products include Box-Jenkins statistical analysis, which gives us the potential for risk analysis on each project—but Seagate Holos does." Seagate Holos also supports Monte Carlo simulation, another commonly used method for risk analysis. Although BT doesn't require that all projects have risk analysis performed on them, some program and project managers need to do it. It's convenient not to have to use a different package, especially since most users see Seagate Holos as an extension to the familiar spreadsheet environment.

B.12.2 Training

Seagate Holos' multidimensionality, however, has meant learning to think in a different way. Where managers used to work with two-dimensional spreadsheets, they can now drill through levels of increasing detail right down to individual transactions. One particularly powerful feature, as BT sees it, is the ability to make forecasts by dragging points on a graph. Seagate Holos automatically changes the underlying figures.

The project's four-person help desk handles both usage and access rights and training. The latter has evolved from a series of half-day hands-on classes for groups of 12 students (some 1,200 Seagate Holos users were "seeded" thanks to this major investment) to self-paced computer-based training (CBT) coupled with application-specific online help files. BT contracted out to have the custom CD-ROM-based CBT made, but provided the material, screen captures, and so on.

B.12.3 More flexible architecture

However, as is often the case, the strengths of Seagate Holos highlighted weaknesses elsewhere. The company's existing database architecture came under the spotlight, and data warehousing seemed the appropriate solution. "We realized that, if we were simply to pull data from our two source systems, we would not generate all the benefits Seagate Holos could deliver," says Gibson. "So we looked at implementing a three-tier system, Oracle data warehouse between the source systems and FIRE."

The new project took 12 months. Meanwhile, a business reorganization brought the D&P group together with Worldwide Networks, forming the Networks and Services (N&S) division, phasing out its self-accounting status, so a further appraisal of the company's information needs was due. "Clearly, it was no use reporting on a division that no longer existed. We needed a data warehouse that dealt with projects on a BT-wide basis," Gibson says. "And based on the experience of the D&P project, we were able to put some designs together."

"We have more than 10,000 projects across N&S, so it had to be a massive system," he says. "One of the great benefits of a data warehouse architecture with a separate reporting application is that it is scaleable. With one or two caveats, the whole architecture design can be scaled."

And so IRIS took root, replacing FIRE as the reporting tool, tracking the costs of projects for more than half of the company. Gradually, more and more datamarts will be developed, obviating the need to rewrite interfaces every time the data in the repository changes. Gibson believes the company's approach breaks new ground. While most datamarts today are created manually, BT is automating the process. Analysts, developers, and customers order information from a warehouse catalogue. The system checks and fulfills these orders and creates the mart automatically. To improve performance, marts may hold pointers to information in the warehouse rather than actually replicating data.

"We don't believe anyone else had gone so far with a data warehouse," Gibson says. "This will really do away with people collecting data separately and then discussing the differences between the numbers, rather than the numbers themselves. Across a business as big as BT, it can make a difference of millions of pounds each year."

IRIS uses two Sun SPARCcenter 2000s to house the project data warehouse, up from the single SPARCserver 1000 used for FIRE. "Sun is well down the road on data warehousing solutions," says Gibson. "Although we are building this solution ourselves, it is clear that Sun understands both the data warehouse architecture and our needs—so they are able to help us as required."

B.12.4 Data cleansing

BT uses SQL Group's Information Junction (IJ) for cleansing the data from its source systems. IJ is a generic interfacing toolset that manages the whole process of transforming and transferring data between different systems. Essentially a highly specialized Oracle7 database application, IJ can talk to any data store. "BT liked the fact that SQL drives the whole transformation process from a relational database model," says Rob Anderson, Professional Services Manager at The SQL Group. "The process is embedded in Oracle and forms the basis of their repository strategy."

Information Junction extracts information, decodes it to remove any proprietary characteristics, applies the necessary conversion rules, and then exports it to the target system. During the transformation, Information Junction performs several operations to ensure the validity of transferred data. Information from one source can be matched with data from other sources, so that it is only transferred if and when the whole transaction is complete. "The advantage of the relational model is that data can be audited down to the finest level of granularity, which allows the end user to monitor and repair data as necessary," Anderson says. Gibson puts it more succinctly, "Data warehousing wouldn't work without IJ."

B.12.5 Business as normal

According to Gibson, data warehousing at BT is now business as normal. From the start, Gibson's prime objective has been to reduce business costs by managing development projects more efficiently. FIRE delivered results almost immediately. Gibson calculates the system saved BT more than four million pounds in the first year it became operational. "Cost saving is extremely important to the P&L," he says. "If you can save four million pounds by implementing the right systems, it is likely that you will save four million every year, because once you have reduced a cost, it stays reduced," he explains. "That is why I see it as so important."

Because it makes so much more information about development projects visible to managers, FIRE has radically altered the way people work, and IRIS will continue to carry through change. "With any new solution today, you have to ask the question, does it change the way people work and does it improve it? If you win commitment from customers, you can use systems as an agent of change," Gibson says. "The response from users indicates the system is enabling them to work in smarter ways."

B.13 Ameritech uses Seagate Holos for sales reporting

Ameritech's Consumer Services division accounts for over one-third of the Chicago-based telecommunications giant's revenues of $13.4 billion (1995). As of early 1997, Consumer Services handled some 13 million access lines in the Midwest five-state region.

Like most regional telecommunications firms facing increased competition for local customers, its "customer-oriented" focus is a lot more than lip service, especially to Katie Boudas, VP of Customer Operations. Early on, Boudas championed the idea of a data warehouse because she saw it as an opportunity to improve customer support. She wanted information about customers' orders to be available throughout the sales channel as soon as the orders were placed, not just when the orders were completed and billed.

B.13.1 Needed: more timely sales reports

The problem was that sales reports simply weren't current enough. Even though Ameritech's 5,000-plus sales agents received sales reports electronically (with much larger monthly paper-based reports going to over 100-plus managers), this didn't give the sales force the up-to-date information they needed to answer questions like, "What's the status of the orders I entered yesterday?" or "How many Caller-ID units have I sold this month in total?"

Until late 1996, Ameritech, like British Telecom, relied on Focus mainframe-based reports that were both hard and expensive to maintain. So, in early 1996, sales management convinced Boudas to fund a new paperless online service tracking (POST) system. With impressive nimbleness, a team was established, and by March, Computer Services Corporation (CSC) was chosen to create an operational system. CSC's efforts and the project were managed by Peter Gamoke, Sr. Manager of Development.

"CSC did the back-end work," recalls John Strobel, an engineer in Ameritech Services' Business Solutions Development unit. They did the analysis (most of the source data remains mainframe-based) and designed the new data warehouse, an Oracle 7.3 database running on a Sun SPARCserver 2000. Within five months, by October 1996, the system was operational, and a beta was launched using a single sales office. All 23 offices were scheduled to go "live" by February 1997. Only then would the existing Focus reports, still running in parallel, be phased out.

The system accommodates between seven and eight million transactions monthly and is updated nightly. The Seagate Holos part of the update takes an average of 1.5 hours nightly, but the Oracle data processing takes between four and five hours. That's because Oracle has to add information from the order entry system and perform several types of reconciliations (the data cleansing). The Oracle data warehouse (about 60 GB worth) keeps data for the previous three months and the current month. According to Strobel, "Every month, we roll off the oldest month to keep the database size manageable so we can update overnight." History, however, is maintained in Seagate Holos. "We tentatively plan to keep one year's worth of Seagate Holos structures online."

B.13.2 Seagate Holos reports

Strobel developed two major reports using Seagate Holos. The main report contains daily and month-to-date quantities by product group. It matches the monthly objective with daily and month-to-date (MTD) revenue, calculates percentage objective achieved, and estimates the percent achieved for month-end based on the current pace of orders. MTD revenue ranking within the office and state is provided, along with the number of orders updated and errored, service order accuracy, and the number of orders that are still pending completion. From this report, a manager can drill into the Oracle database to check on status of orders, any errors that have been generated, as well as order tallies for the individual sales reps. This report can be viewed for any day of the month, and for current as well as previous months, and is consolidated from sales reps to team, office, state, and regional totals.

The second report contains month-to-date gross quantities and revenue, adjustments, and net quantity and revenue by product code. Quantities and revenue allow drill-down to show the order type, and adjustments allow drill-down to show the adjustment reason. This monthly report can be viewed for current as well as previous months and follows a similar consolidation scheme. Prior to POST rollout, users were given training, with at least five individuals from each sales office's customer care center attending two-day POST system classes. Seagate Holos, the main "front-end" analysis tool, is Ameritech's standard multidimensional analysis engine. Because the Seagate Holos component is quite intuitive, managers were taught how to use it in about half a day.

According to Strobel, "The security for our system is efficient. The sales reps—there are approximately 5,000 of them—have access to view their own results only (exactly like the Seagate Holos report) through a simple Visual Basic application." (Visual Basic is also used to update static data such as data about the sales reps themselves and product tables.) However, the 400 or so managers who use Seagate Holos and attended training have access to all of their own sales reps' information and can also view results from other offices and channels, but only at the office level of detail. As far as upper management goes, "VPs and directors generally have little interest in drilling down to individual rep performance," according to Strobel, so they have access to summary data for all offices in all channels. Administrative staff have access to everything (all reps in all offices and channels).

B.13.3 Futures

We interviewed Strobel and his boss, David Smith, about a year after POST was first conceived, when POST was being rolled out into the final sales offices, so there weren't any quantitative measurements of costs, savings, or increased customer satisfaction yet. Managers were reporting, however, that they had been using the new reports for coaching. The combination of drill-down and more timely information made it easy to identify not only sales reps who weren't selling well, but also reps who preferred selling certain products, and so on.

Strobel and Smith already have plans for additional integration. And the fact that upper management has put in requests for more summary data is a sure vote of confidence. What's on the agenda? Improved integration (and faster processing) with the feeder systems, tie-ins with call-volume management, and tie-ins with the sales force scheduling system.

Strobel and Smith's customers for the POST system are the Ameritech sales representatives and their managers. Since those folks' goals are to make sales, POST has to be responsive to them. But Strobel and Smith also want to help POST's early champion, Customer Service VP Boudas, with her goal of increasing customer satisfaction. We were impressed with both their delivery and with Ameritech's vision. It shows how data warehousing can be used effectively for sales force management. Integration with a departmental EIS is surely coming. After all, at press time, Strobel and Smith were just completing the initial rollout to the last sales offices and had already received interest in the new system from Consumer Marketing staff.

Appendix C

SPSS

Become more competitive and productive by empowering your knowledge workers with analytical tools on their desktop to perform more extensive data analysis such as:

- Finding the most profitable, successful markets and customers to grow revenue;
- Segmenting your customers for more cost-effective targeted marketing programs;
- Forecasting sales more accurately for better revenue management.

C.1 Extend your analysis

Statistics are essential for data mining. Statistics enable you to mine your data in ways not possible with spreadsheets, report writers, and query tools. Data mining with statistics complements and supplements the analysis you probably do today.

Use statistics to get better information on which to base your decisions. You can uncover hidden relationships, create predictive models, find clusters in your database, discover associations between activities, and review deviations. You can react more quickly to opportunities and threats with SPSS because it helps you find the information you need quickly and easily.

With SPSS, you can easily extend your analysis within a comprehensive environment. SPSS is an open system, so you can assemble a "best of breed" data warehousing and data mining solution comprised of the best software in each category. SPSS partners with other data mining vendors to provide you with easy-to-use interfaces between products. This approach gives you the most powerful and appropriate tools for your data warehousing and data mining efforts.

C.2 Use a comprehensive solution

SPSS delivers each of the four A's of data mining: access, analysis, action, and automation. No other vendor provides a comprehensive analysis solution delivering both verification and discovery techniques to help you get a return on your investment on the large amounts of data you have collected:

Access: With SPSS, you can easily access your data wherever they reside—data warehouse, data mart, databases, or popular spreadsheet files. You can also enter data directly into SPSS and join data sets from different sources. An easy-to-use wizard reads popular spreadsheet and database files and prompts you step-by-step through joining multiple database tables so you can work with more of your data. SPSS also includes direct links to many query and reporting and OLAP tools that provide simultaneous access to multiple databases. And, SPSS reads files up to terabytes in size.

Analysis: In data mining, analysis includes both verification techniques (traditional statistics) and discovery techniques (such a neural networks). SPSS provides both. SPSS has multidimensional tables, basic and advanced regression, forecasting methods, correlation and cluster analysis, and many other statistical procedures for verification. SPSS offers a product of the popular trees-based CHAID algorithm, and neural network and data visualization products for discovery.

Action: SPSS facilitates you acting on your findings. Since your analysis is based on statistics, you can be more confident in your decisions. Your findings can be easily embedded into a presentation, an e-mail, or a report. SPSS offers you the flexibility to answer follow-up questions without requiring the information system (IS) to create a new view of the data or a new report. This ad hoc analysis capability helps you make decisions faster. Results can easily be shared by sending flexible pivot tables to others so they can further explore the results on their own.

Automation: SPSS provides a fast way for analytical capabilities to be delivered to a wider audience by offering the ability to put sophisticated analysis behind a "push button" interface.

C.3 Feel confident in your choice

Rely on nearly 30 years of statistical experience to help you mine your data. Experienced SPSS analysts provide comprehensive beginning to advanced level training in product operations, statistical methods, and applied statistics on a regular basis in cities around the world. SPSS offers complete consulting services around the world for all your data mining needs.

C.4 Company information

SPSS provides a range of products for people with varying levels of statistical expertise. Count on a proven, dependable software tool used by more than 100,000 customers around the world. SPSS is the leader in desktop statistical software and has won several awards including the prestigious Win 100 award from *Windows Magazine*. You will work with the industry standard because SPSS has received Microsoft's certification for Windows NT and Windows 95 software.

SPSS' headquarters are in Chicago, Illinois. SPSS is available from offices and distributors around the world. To place an order or to get more information, contact your nearest SPSS office or visit our World Wide Web site at http://www.spss.com

C.5 SPSS offices

SPSS Inc.
United States and Canada
444 North Michigan Avenue
Chicago, IL 60611
Telephone: +1.312.329.2400
Toll-free: +1.800.543.2185
Fax: +1.312.329.3668

SPSS Federal Systems (U.S.)
Courthouse Place
2000 North 14th, Suite 320
Arlington, VA 22201
Telephone: +1.703.527.6777
Toll-free: +1.800.860.5762
Fax: +1.703.527.6866
E-mail: corinne@spss.com

SPSS Argentina srl.
Piso 8, A, Arenales 1140
Buenos Aires (1061)
Argentina
Telephone: +541.816.4086
Fax: +541.814.5030
E-mail: info@spss.com.ar

SPSS Asia Pacific Pte. Ltd.
70 Bendemeer Road
#04-01 Hiap Huat House
Singapore 339940
Telephone: +65.3922.738
Fax: +65.3922.739
E-mail: cecilialim@acapacific.com.sg

SPSS Australasia Pty. Ltd.
121 Walker Street, North
Sydney, NSW 2060
Australia
Telephone: +61.2.9954.5660
Toll-free: +1800.024.836
Fax: +61.2.9954.5616

SPSS Israel Ltd.
Herzlia Business Park
8 Maskit St.
Herzlia 46733
Israel
Telephone: +972.9.526700
Fax: +972.9.526715
E-mail: ilant@ibm.net

SPSS Italia srl
Via Ciamician, 4
I - 40127 Bologna
Italy
Telephone: +39.51.252573
Fax: +39 51.253285
E-mail: nicola@spss.it

SPSS Japan Inc.
2-2-22 Jingu-mae
Shibuya-ku,
Tokyo 150
Japan
Telephone: +81.3.5474.0341
Fax: +81.3.5474.2678E-mail: noguchi@spss.co.jp

SPSS Korea
#201 Chunil Building
826-26 Yeoksam-Dong
Kangnam-KU
Seoul
Korea
Telephone: +82.2.552.9415
Fax: +82.2.539.0136
E-mail: spsskic@nuri.net

SPSS Latin America
444 North Michigan Avenue
Chicago, IL 60611
Telephone: +1.312.494.3226
Fax: +1.312.494.3227
E-mail: laurab@spss.com

Appendix C: SPSS 243

SPSS Belgium
Naamse steenweg 62
3001 Heverlee
Belgium
Telephone: +32.162.389.82
Fax: +32.1620.0888
E-mail: ignace@spss.be

SPSS Benelux BV
P.O. Box 115
4200 AC Gorinchem
The Netherlands
Telephone: +31.183.636711
Fax: +31.183.635839
E-mail: wiepke@spss.com

SPSS Central and Eastern Europe
c/o SPSS UK Ltd.
1st Floor, St. Andrew's House
West Street
Woking, Surrey GU21 1EB
United Kingdom
Telephone: +44.(0)1483.719200
Fax: +44.(0)1483.719290
E-mail:
100525.1660@compuserve.com

SPSS East Mediterranean & Africa
Herzlia Business Park
8 Maskit St.
Herzelia 46733
Israel
Telephone: +972.9.526700
Fax: +972.9.526715
E-mail: tal.spss@formula.co.il

SPSS France SARL
72-74 Avenue Edouard VAILLANT
92100 Boulogne
France
Telephone: +33.1.4699.9670
Fax: +33.1.4684.0180
E-mail: kpilmann@spss.fr

SPSS Malaysia Sdn Bhd
4th Fl., Block E, Kelana ParkView Tower
Jalan SS 6/2, Kelana Jaya,
47301 Petaling Jaya,
Selangor
Malaysia
Telephone: +603.704.5877
Fax: +603.704.5790
E-mail: carol@acapacm.po.my

SPSS Mexico Sa de CV
San Lorenzo 153
Despacho 1101
Col. del Valle
Mexico D.F. 03110
Mexico
Telephone: +52.5.575.3091
Fax: +52.5.575.3094
E-mail: mexico@spss.com

SPSS Middle East and South Asia
c/o Aptec UAE
Suite M07, Juma Al Majid Building
Bank Road
PO Box 33550
Dubai
United Arab Emirates
Telephone: +971.4.525536
Fax: +971.4.524669
E-mail: 101326.2301@compuserve.com

SPSS Scandinavia AB
Gamla Brogatan 36-38, 4th floor
111 20 Stockholm
Sweden
Telephone: +46.8.102610
Fax: +46.8.102550
E-mail: josefson@spss.se

SPSS Schweiz AG
Seestrasse 11
8027 Zurich
Telephone: +41.1.201.0930
Fax: +41.1.201.0921
E-mail: 100102.1236@compuserve.com

SPSS Germany
Rosenheimer Straße 30
D-81669 München
Germany
Telephone: +49.89.4890740
Fax: +49.89.4483115
E-mail: mluehe@spss.de

SPSS Hellas SA
5 Ventiri Street
115 28 Athens
Greece
Telephone: +30.1.7251925 / 7251950
Fax. +30.1.7249124
E-mail: orcospss@mail.hol.gr

SPSS Hispanoportuguesa S.L.
C/ LUCHANA, 23 5ª PLTA.
28010 Madrid
Spain
Telephone: +34.1.447.37.00
Fax: +34.1.448.66.92
E-mail: 100136.2666@CompuServe.COM

SPSS Ireland
Clifton House
Lower Fitzwilliam Street
Dublin 2

SPSS Singapore Pte. Ltd.
70 Bendemeer Road
#04-01 Hiap Huat House
Singapore 339940
Telephone: +65.2991238
Fax: +65.2990849
E-mail: cecilialim@acapacific.com.sg

SPSS Taiwan Corp.
12th Fl., No. 126, Nanking East Road, Sec 4
Taipei, Taiwan
Republic of China
Telephone: +886.2.5771100
Fax: +886.2.5701717
E-mail: a001@ms1.hinet.net

SPSS UK Ltd.
1st Floor, St. Andrew's House
West Street
Woking, Surrey GU21 1EB
United Kingdom
Telephone: +44.1483.719200
Fax: +44.1483.719290
E-mail: rachel@spss.co.uk

Appendix D

The DecisionWORKS suite from Advanced Software Applications

THE DecisionWORKS SUITE is a set of integrated products for automated data mining, data analysis, and decision support. They encompass an array of analytical outcomes for marketing professionals, including predictive modeling, custom clustering, profiling, and segmentation. But the most intriguing aspect of this product suite is its appeal to business users. Because all of the underlying mathematical functions are automated, and all of the necessary safeguards are designed into the software, the user does not have to be statistically or technically oriented. As long as you know your data and how you would like use that data to better address your business needs, you can use DecisionWORKS. ModelMAX and dbPROFILE are key products within this product suite, and are highlighted on the following pages.

For more information, contact A.S.A. directly at:

Phone 412-429-1003
Fax 412-429-0709
e-mail asa@fyi.net
Mail 333 Baldwin Road
Pittsburgh, PA 15205

D.1 ModelMAX: the new standard for predictive modeling

Since its introduction in 1993, ModelMAX has become the top-selling predictive modeling software. As the only truly automated package of its type, its leading position shouldn't come as a surprise. ModelMAX is specifically designed for marketers who want to enjoy the many benefits of predictive modeling, regardless of their technical expertise. Based on a hybrid of advanced technologies including a neural network, it quickly analyzes data to identify individual patters of behavior. These patterns reveal the likelihood of someone behaving in a certain way, such as responding to a promotion or remaining a single-time buyer versus becoming a long-term or high-volume buyer. They can also predict complex outcomes, such as the estimated lifetime value of a customer. ModelMAX continues to be a breakthrough for marketers because of its speed, ease of use, and reliability in making these ever-increasingly important predictions. ModelMAX gives small and medium-sized organizations an affordable way to implement modeling, and allows large-sized organizations to quickly expand how much and how fast they model. This incredible business-oriented PC software package is scaleable and integrates into virtually any environment. And it doesn't matter if your promotion is via the mail or the phone. You save significant amounts of time and money while boosting revenue and profits.

D.1.1 Software for today's marketer

Whether you are looking to acquire, develop, or retain customers, ModelMAX can help you realize the full power of your database and the full promise of database marketing. Plus, by modeling more frequently, you can better stay on top of your market and your customers' changing needs. ModelMAX is a tremendous boost to all marketers who want to do their own modeling. It is also a major productivity tool for statisticians who want to complete more analytical projects in less time. Predictive modeling is especially helpful for telecommunications. You can predict who is most likely to purchase additional or premium services, as well as who is most likely to switch to the competition for long-distance service or other competitive products. US West and MCI are just two examples of a long list of companies worldwide that use ModelMAX as an integral part of their marketing activities.

The whole idea behind targeted marketing is to reach a better qualified audience. Predictive modeling is a key component to that strategy. But for most marketers, predictive modeling has been cost-prohibitive and fairly difficult to implement. Thanks to ModelMAX, these barriers no longer exist. Now you can efficiently and cost-effectively select the optimum model for every promotion and marketing project.

D.1.2 Simple yet powerful

ModelMAX is truly unique because it brings the power of predictive modeling to you, the marketing professional. It is an end-user application featuring standard marketing terms; point-and-click functionality; fast, reliable results; and extensive management reports.

ModelMAX doesn't require you to be an expert in statistics or to understand the intricacies of advanced technologies and complicated mathematical algorithms. It doesn't require special expensive hardware. All ModelMAX requires is that you have a basic business-level understanding of your market, customers, and prospects. If you know your data and what you would like to do with it, ModelMAX is for you. Here are just a few of the different types of predictive models you can create with ModelMAX:

- Differentiate between prospective buyers and nonbuyers;
- Distinguish single-time buyers from multitime buyers;
- Rank order buyers based on future purchases, profits, or donations;
- Estimate how much someone will spend or how many units they might buy;
- Forecast when the optimum time is to promote to someone;
- Improve customer retention and lower your attrition rates;
- Uncover cross-sell and up-sell opportunities.

D.1.3 The advantages of advanced technologies

ModelMAX is based on a hybrid of advanced technologies and statistics. The primary technology is a neural network, a powerful nonlinear modeling technique that is capable of adaptive learning and identifying patterns in customer and market-related data.

Because ModelMAX identifies patterns, not trends or central tendencies, the models that it generates are more robust and the predictions hold truer than those generated by traditional methods. With ModelMAX, you no longer have to settle simply for segmenting your files into like categories. Instead, ModelMAX selects the individual records that most closely match your desired outcome, letting you execute the type of targeted promotions that are so vital to your business. There may be many different patterns in an audience, and ModelMAX will find them all.

Unlike statistical packages and other advanced technology approaches that place a technical burden on the user, ModelMAX shields you from a steep learning curve. We've automated the mathematics so you can concentrate on results, not the mathematics.

D.1.4 Combine predictive power with descriptive insight

ModelMAX's predictive modeling and scoring capabilities make it a natural companion to dbPROFILE, A.S.A.'s custom clustering and segmentation tool. ModelMAX is seamlessly integrated with dbPROFILE via the Direct Data Xchange (DDX) feature. In addition to linking these two products together, DDX also gives you easy access to your data. DDX enables you to access open database connectivity (ODBC)-compliant databases as well as dbf formats, so you can automatically extract data from a wide range of popular databases, including

- Sybase;
- Oracle;
- dBASE;
- Access;
- Paradox;
- Informix;
- Foxpro.

DDX also makes it very convenient to model and cluster on the same project file. For example, before you create a model with ModelMAX, dbPROFILE helps you determine which variables are related to a certain behavior. By narrowing in on only those variables that are relevant, you save model-building time and disk space. Once a model has been built, dbPROFILE helps you determine how each segment is unique. You can accurately view ModelMAX segments via cross-tabs, queries, and most importantly, perception maps that actually display the natural groupings found in the data.

D.2 dbPROFILE: a breakthrough for custom clustering and data visualization

dbPROFILE is an affordable, automated custom clustering and data visualization tool for marketing professionals in all fields, including research and analysis, communications, product management, business development, and direct marketing. A powerful, Windows-based solution, dbPROFILE offers an easy-to-use method for visualizing the inter-relationships of virtually any type of data. Study demographic, geographic transactional, behavioral, lifestyle, survey, product, and related data in new ways that enhance your understanding of customer bases and markets. Based on a hybrid of statistics and advanced technologies, including a neural network, dbPROFILE clusters and segments quickly and effectively to help you analyze the qualitative aspects of your data to support a wide range of business decisions. Results can be far ranging, including being able to anticipate product affinities, fine-tune your media mix, prioritize list selects, gauge the effectiveness of marketing communications, learn the impact of special situations such as price sensitivities, and track a wide range of other marketing activities. dbPROFILE's special design and breakthrough technology offers a cost-effective alternative that gives you the power to perform dynamic data discovery and data mining at the convenience of your PC. dbPROFILE opens up a whole new world of analytical insight that quite simply was not possible before. With dbPROFILE, you really are only limited by your imagination.

D.2.1 Unlock the mysteries of your database

The ultimate goal of any targeted marketing effort is to reach the most qualified, motivated audience with the right products and messages at the right time. Customer clustering helps you derive the information you need to ensure your marketing efforts are as refined and successful as possible:

- Custom clusters are effective for identifying key attributes and determining who groups together and why. With custom clustering, you can exploit the natural descriptive power of your data to its fullest.
- Identify customer characteristics based on virtually any combination of variables, including individual behavior or transactional data.

- Leverage your database more effectively by organizing it into actionable segments and like groups.
- Assign unique cluster codes to each customer in your database and communicate with each group according to its attributes.
- Use cluster codes in other analytical processes, such as predictive modeling with ModelMAX, to pinpoint those people most likely to respond or to purchase.
- Track and measure your customer's behavior over time and gauge the effectiveness of your marketing efforts, even at a micro level.

More specifically, you can acquire new customers more easily by defining and prioritizing selects that guide list rentals. Target existing customers for specialty or loyalty marketing programs. Identify key attributes and target marketing messages to your audience's specific needs or desires. Understand what influences a product's acceptance or development. Sharpen the focus and effectiveness of your advertising. And identify segments that warrant additional research.

Many marketers are familiar with cluster codes that are largely based on census or lifestyle data. While this type of data can be valuable, it cannot deliver the same unique insight as performing customer cluster analysis on your own data. In fact, we encourage our clients to consider including demographic and lifestyle data when they use dbPROFILE. These additional elements can help strengthen your analysis and make the process even more valuable.

D.3 For more information

If you would like to learn more information about ModelMAX, dbPROFILE, or any other products under our DecisionWORKS suite for integrated data mining, data analysis, and decision support, contact A.S.A. at 412-429-1003. We'll be happy to discuss your specific needs, send you information, and even arrange a free evaluation of our software if you like.

Glossary

Data management terms

Aggregation A summarization of data to a higher level. (i.e., A geographical aggregate might show the total of all sales for a given country. The sales data is aggregated to the country level.)

CIO Chief information officer—The highest level executive within an organization, responsible for the management of the company's computer systems and information.

CKO Chief knowledge officer—The highest level executive within an organization, responsible for the management of all knowledge within the organization.

Client-server application Technical architecture of an information system, consisting of large machines that service a large number of users (servers) and small machines that end users manipulate to do real work (clients).

Customized applications Software—Applications or software only dedicated to certain domains or applications.

Data administrator The person responsible for the management of all of the data within a corporation's environment.

Database management The management of large-scale database systems.

Data briefing Type of report including key numbers and key target values.

Data discovery Operations or treatments allowing for the discovery of new meaning or information contained within existing data stores.

Data mining Operations or treatments involving data treatments, whether they are numerical, statistical, or textual that drill into existing data stores, attempting to discover new meanings or insights.

Data model Abstract representation of database organization.

Data reporting Type of function associated with summarizing significant data.

Data supervision Data treatments associated with constraints of time.

Data warehouse Other term for information system or information warehouse.

Data workstation Equipment and software used by end user in order to gain access to the warehouse.

Decision support system (DSS) Software associated with the decision-making process.

Dimension Characteristic of an OLAP application that defines one of the aggregation and navigation characteristics. A dimension defines the data characteristics that people can use to aggregate to.

Executive information system Software associated with decision-making process by executive managers. Another term is enterprise information system.

Functional architecture   Architecture describing the main functionalities of an information system only.

Geographical information systems (GIS)   A system that codifies numerical data and displays it in the form of a geographical map.

Mainframe   Any of a class of large-scale computer systems that run many corporate data centers.

Metadata   Information about information. Metadata describes the overriding characteristics of the data itself (i.e., field names and file names are forms of metadata).

On-demand analysis   Analysis that is executed when the user explicitly requests it.

OLAP   Online analytical processing—Any of a class of software products that features a functionality consisting of a spreadsheet-like front end and the ability to navigate across many dimensions simultaneously.

Operational system   Frontline system recording business operations.

Prototyping   Technique used to test the feasibility of the assumptions made about a proposed system's performance by building and testing a small portion of its overall functionality.

Querying   Software able to request data from a distant station.

Relational database   Type of database used to support operational and data warehouse systems.

Server   Station or workstation containing data of an information system based on a client-server application.

Spreadsheet   Software able to manipulate data from a data matrix.

Technical architecture   Architecture describing the technical features of an information system (hardware, software, network tools, protocols, etc.).

UNIX Operating system common to most "open systems" computer environments.

User information system Software associated with the decision-making process by any type of end user for any type of application (exploratory, commercial, financial, quality, marketing, business).

Data analysis terms

Bivariate analysis Method that highlights main features of relationships between two variables (quantitative or categorical). This method involves scatter plots, data cross-tables, mosaic diagrams, and three-dimensional histograms.

Categorical variable A categorical variable simply records into which of several categories a person or thing falls.

CHAID Chi-square analysis measures the observed frequencies of values for a categorical value and determines how well they correspond. CHAID is used to discover statistically significant groupings of a population.

Cluster analysis Method that build groups, subgroups, nested groups, hierarchical groups of cases, or variables, often represented by a dendogram.

Correlation analysis Method that provide insights into the relationships between variables based on how similarly they carry different values for different variables.

Discriminate analysis Method that projects relationships between two or several groups of cases, represented as scatter plots.

Distribution Pattern of variation of a variable. The distribution records numerical values of a variable and how often each value occurs.

Exploratory data analysis Method that highlights main features of relationships between several variables (multiscatter plots, multiways data sets, sunrays plots, profiles diagram).

Exploratory variable Variable expected to explain the observed outcome.

Indicators   Key variable resulting from several variables, or several indicators, often associated with key target values.

Linear regression   Method that proposes a linear modeling of data. The linear regression may be simple (involving two variables), or multiple (involving several variables). There are also nonlinear regression methods.

Multiple correspondence analysis   Generalization of simple correspondence analysis to several categorical variables. Method that projects relationships between several categorical variables onto successive scatter plots as well as the cases associated with these variables.

Multidimensional scaling   Method that projects relationships between cases. Scatter-plot relationships given by a distance matrix between the cases.

Neural network   A software program designed to discover hidden relationships between members of a data population that uses neural programming; a programming technique that mimics the workings of the human neurological system.

Principle components analysis   Method that projects relationships between variables onto scatter plots. This method, associated with the linear coefficient of correlation, also projects cases associated with variables.

Quantitative variable   A quantitative variable takes numerical values for which arithmetic operations such as differences and averages make sense.

Response variable   A response variable measures an outcome of a study.

Simple correspondence analysis   Method that projects relationships between two categorical variables onto successive scatter plots.

Statistical tests   Computations able to identify if a relationship between two objects is true or not according to a certain confidence percentage.

Target indicators   Key variable or indicator, fixed by the end user, whose values must be reached after a certain period of time.

Univariate data analysis   Method that highlights the main features of a variable (histograms, stem plots, box plots, pie charts, bar charts, components bar charts, time series graphs).

Variable   Any characteristic of a person or thing that can be expressed as a number is called a variable.

Verbatim analysis   Analysis of textual data.

Selected bibliography

Anderberg, M. R., *Cluster Analysis for Applications*, New York: Academic Press, 1973.

Barquin, R. C., and H. A. Edelstein, *Planning and Designing the Data Warehouse*, Englewood Cliffs, NJ: Prentice-Hall, 1996.

Becker, R. A., and J. M. Chambers, *An Interactive Environment for Data Analysis and Graphics*, Wadsworth, CA, 1984.

Benzecri, J. P., *Correspondence Analysis Handbook*, M. Dekker Inc., 1992.

Blyth, J. W., and M. M. Blyth, *Telecommunications: Concepts, Development and Management*, Bobbs Merril Educational Publishing, 1984.

Canavos, G. C., *Applied Probability and Statistical Methods*, Boston, MA: Little, Brown and Co., 1984.

Chambers, J. M., *Computational Methods for Data Analysis*, New York: John Wiley & Sons, 1977.

Chambers, J. M., et al., *Graphical Methods for Data Analysis*, Boston, MA: Duxbury Press, 1983.

Cleveland, W. S., *Dynamic Graphics for Statistics*, Wadworth and Brooks/Cole, 1988.

Cleveland, W. S., *Visualizing Data*, Summit, NJ: Hobart Press, 1993.

Coxon, A.P.M., *The User's Guide to Multidimensional Scaling*, Heineman Educational Books, HEP Paperback, 1982.

Deming, E. W., *Out of the Crisis*, Cambridge, MA: MIT Press, Center of Advanced Engineering Study, 1986.

Elbert, B. R., *International Telecommunications Management*, Norwood, MA: Artech House, 1990

Feigenbaum, A. V., *Total Quality Control*, New York: McGraw-Hill, 1983.

Gnanadesikian, R., *Methods for Statistical Data Analysis of Multivariate Observations*, New York: John Wiley & Sons, 1977.

Greenacre, M., *Theory and Applications of Correspondence Analysis*, London: Academic Press, 1984.

Hartigan, C., *Clustering Algorithms*, New York: John Wiley & Sons, 1975.

Hausman, J. A., R. L. Nolan, and S. P. Bradley, *Globalization, Technology, and Competition: The Fusion of Computers and Telecommunications in the 1990s*, Cambridge, MA: Harvard Business School Press, 1993.

Hayashi, C., et al., *Recent Developments in Clustering and Data Analysis*, London: Academic Press, 1988.

Heldman, R., *Future Telecommunications: Information Applications, Services & Infrastructure*, New York: McGraw-Hill, 1993.

Hudson, H., *A Bibliography of Telecommunications and Socio-Economic Development*, Norwood, MA: Artech House, 1988.

Inmon, W., and R. Hackathorn, *Using the Data Warehouse*, New York: John Wiley & Sons, 1994.

Inmon, W., *Building the Data Warehouse*, QED, 1990.

Ishikawa, K., *What Is Total Quality Control?*, Englewood Cliffs, NJ: Prentice Hall, 1985.

Jambu, M., and M. O. Lebeaux, *Clustering Analysis for Data Analysis*, North Holland, Amsterdam, the Netherlands, 1983.

Jambu, M., *Exploratory and Multivariate Data Analysis*, New York: Academic Press, 1992.

Jardine, N., and R. Sibson, *Mathematical Taxonomy*, New York: John Wiley & Sons, 1971.

Kendall, M. G., and A. Stuart, *The Advanced Theory of Statistics*, London: Griffen, 1973.

Kachigan, S. K., *Statistical Analysis*, New York: Radius Press, 1986.

Kelly, S., *Data Warehousing: The Route to Mass Customization*, New York: John Wiley & Sons, 1994.

King, D. R., and R. W. Blanning, *Current Research in Decision Support Technology*, IEEE Computer Society Press, 1993.

Kruskal, J. B., and M. Wish, *Multidimensional Scaling*, Beverly Hills, CA: Sage Publications, 1978.

Laska, R., and A. Paller, *The EIS Book: Information Systems for Top Managers*, Irwin Professional Pub Computer, 1990.

Lebow, I., *Information Highways and Byways: From the Telegraph to the 21st Century*, IEEE Computer Society Press, 1995.

Mattison, R., *Data Warehousing: Strategies, Technologies and Techniques*, New York: McGraw-Hill, 1996.

Mattison, R., *Understanding Database Management Systems*, New York: McGraw-Hill, 1987.

Moore, D. and G. P. McCabe, *Introduction to the Practice of Statistics*, San Francisco, CA: Freemann, 1983.

Mueller, M., and Z. Tan, *China in the Information Age: Telecommunications and the Dilemmas of Reform*, Praeger Pub Text, 1996.

Nishisato, S., *Analysis of Categorical Data: Dual Scaling and its Applications*, Toronto: University of Toronto Press, 1986.

Petrazzini, B. A., *The Political Economy of Telecommunications Reform in Developing Countries: Privatization and Liberalization in Comparative Perspective*, Praeger Publishing, 1995.

Plantamura, V. L., G. Visaggio, and S. Branko, *Frontier Decision Support Concepts: Help Desk, Learning, Fuzzy Diagnoses, Quality, Prediction*, New York: John Wiley & Sons, 1994.

Reeves, L. L., and V. Poe, *Building a Data Warehouse for Decision Support*, Englewood Cliffs, NJ: Prentice-Hall, 1995.

Rhodes, P. C., *Decision Support Systems: Theory and Practice*, Alfred Waller Ltd., 1993.

Ryan, T. P., *Statistical Methods for Quality Improvement*, New York: John Wiley & Sons, 1989.

Sappington, D. E., and D. L. Weisman, *Designing Incentive Regulation for the Telecommunications Industry*, Cambridge, MA: MIT Press, 1996.

Schiffmann, S. S., M. L. Reynolds, and F. W. Young, *Introduction to Multidimensional Scaling*, New York: Academic Press, 1981.

Sokal, R. R., and P. H. A. Sneat, *Principles of Numerical Taxonomy*, San Francisco, CA: Freeman, 1977.

Teske, P., (Ed.), *American Regulatory Federalism and Telecommunications Infrastructure*, Lawrence Erlbaum Assoc., 1995.

Tukey, J. W., *Exploratory Data Analysis*, Reading, MA: Addison-Wesley Publishing Co., 1977.

Turban, E., *Decision Support and Expert Systems: Management Support Systems*, Englewood Cliffs, NJ: Prentice-Hall, 1995.

Walford, R. B., *Information Networks: A Design and Implementation Methodology*, Reading, MA: Addison-Wesley Publishing Co., 1990.

Wright, D., *Broadband: Business Services, Technologies, and Strategic Impact*, Norwood, MA: Artech House, 1993.

Zachman, J. A, J. G. Geiger, and W. H. Inmon, *Data Stores, Data Warehousing and the Zachman Framework*, New York: McGraw-Hill, 1997.

About the author

ROB MATTISON is a leading international authority in the areas of database implementation, data warehousing, and data mining in the telecommunications industry. He has authored five popular books in this area to date and maintains an aggressive schedule of speaking, consulting, and writing engagements that keep him on the leading edge of changes in the industry. He can be reached at any time and welcomes comments, questions and conversation through his email address 70451.207@compuserve.com

Index

Access, 88
    area, 84
    SPSS and, 240
    tool types, 88
Accounting. *See* Finance and accounting
Acquisition, 85–86
    area, 84
    job streams, 86
    of multiple platforms, 86
    process of, 85
    right of service, 54–55
    See also Data warehouse environment; Telecommunications systems
Activation, 57–58
    operational efficiency, 109
    value propositions, 104
    See also Telecommunications systems

Agents, 18–19, 88, 97
    defined, 18–19
    enterprise intelligence systems, 219
Air/marine services, 4
Alerts (PowerPlay), 125–26
Alignment
    computer system, 72–74
    information system, 71–72
    organizational, 39, 50
    problems, 43, 74
    symptoms, 74–76
    value chain, 68–71
Ameritech, 235–37
    Consumer Services division, 235
    paperless online service tracking (POST), 235–37
    sales report problem, 235–36

Ameritech (continued)
  Seagate Holos reports, 236–37
  See also Seagate Holos
Analysis of variance (ANOVA), 132
Analytical mining, 130–31
Application layers, 89, 90
AreaCodeInfo, 157

Basic trade area (BTA), 164–65
Billing data
  by boxplot and interaction tools, 186
  polar diagram of, 186
  visualization, 184–85
  See also STATlab
Billing systems, 58–59
  changing, 76
  functions of, 75
  management, 58
  operational efficiency, 109
  "shared," 36
  value propositions, 104
  See also Telecommunications systems
British Telecom, 229–35
  computer-based training (CBT), 232
  data cleansing, 234
  flexible architecture, 233–34
  Flexible Interactive Reporting Environment (FIRE), 231–34
  Information Builders Focus product, 230
  Information Junction (IJ), 234
  Interactive and Reporting Information System (IRIS), 230, 233
  project rationale, 230–32
  training, 232
  See also Seagate Holos
Broadband
  defined, 8
  services, 4
Building
  all-at-once advantages, 80
  cost-justified approach, 89–92
  "dump and run" model, 81–83

summary, 92–93
See also Data warehouses
Business
  20th century theory of, 27
  acquiring right to do, 54–55
  core systems in, 29
  "customer is king" orientation, 10–11
  focusing on, 214–15
  intelligence (*see* Enterprise business intelligence)
  as knowledge transformation, 28
  operations, 62
  strategic options, 10
  technological innovation and, 9
  of telecommunications, 2
Business units
  allocating to value chain, 66–78
  consumer category model, 68
  geographical model, 68
  line of business-based, 68
  list of, 67
  profit center, 68
  special, creation of, 68

CableInfo, 157
CART, 144
Cellsite analysis, 159–62
Cellular, 4
CellularInfo, 157
CHAID, 144, 240
Channel management warehouses, 101
CHURN/CPS, 187–96
  components, 191
  defined, 189
  eight-step guided analysis, 193
  function of, 189
  guided analysis module, 189, 190
  implementation cycle, 193
  implementation steps, 194
  implementing, 191–93
  payback types, 190
  preventive CPS, 190
  proactive CPS, 190–91

user profiles, 194–96
using, 193–96
value of, 189–91
Cluster analysis, 134–35
results, 135
in SPSS, 134
See also Descriptive statistics
COGNOS PowerPlay. *See* PowerPlay
Coin phones, 3
Computer systems
alignment of, 72–74
business functions and, 29–30
customer service, 60
functions of, 210
inventory of, 22
value chain alignment, 72–74
See also Information systems
Consumer category models, 68
ContourInfo, 157
Cost-justified approach, 89–92
application layers, 89, 90
defined, 89
development/construction and, 92
organization/prioritization and, 91–92
value propositions and, 90–92
vision development and, 90–91
Creation
activities, 54
value propositions, 103–4
See also Telecommunications systems
Credit
management, 61–62
value propositions, 99–100
CSQlab, 206–7
control charts, 207
defined, 206
environment, 206
See also STATlab
Customer-based warehouses, 99
Customer intimacy, 108, 129–41
Customer Profile System (CPS), 187
preventive, 190
proactive, 190–91

See also CHURN/CPS
Customers
activity of, 34
contacts, tracking, 33–34
as "king," 10–11
knowledge of, 11
relationship with, 61
Customer service, 59–61
call routing and scheduling system, 103
calls, 60
computer systems, 60
departments, 60
geographical information and, 158
operational efficiency, 109
value propositions, 100
See also Telecommunications systems
Customer services
activating, 57–58
setting up, 57

Data accessibility, 179–81
Data analysis glossary terms, 254–56
DATAboard, 183–84, 187, 207–8
calculations, 208
defined, 207
reporting objects, 207
See also STATlab
Data discovery tools, 19, 88, 98
defined, 19
enterprise business intelligence, 220
DATAman, 203
Data management, 16–17
glossary terms, 251–54
from knowledge point of view, 31
Data mining, 13, 18–20
agents, 18–19
analytical, 130–31
categories, 18–19
data discovery, 19
field of, 18
OLAP, 19
physical implementation of, 40

Data mining (continued)
　query/reporting tools, 19
　Seagate Holos, 228
　statistical analysis tools, 19
　statistics and, 239
　value delivery and, 96–98
　visualization, 19
　Web-based discovery, 19
DATAread, 203–5
　defined, 203
　query results, 204
　See also STATlab
Data redundancy, 16, 17
Data reporting, 185–87
Data transport services, 3
Data warehouse environment
　access area, 84, 88
　acquisition area, 84, 85–86
　functional components of, 84–89
　operational infrastructure, 84
　physical infrastructure, 84
　storage area, 84, 87
Data warehouses
　alternative, 37
　architecture, 39
　building, 79–93
　channel management, 101
　construction of, 37
　cross-silo, 34
　data basements vs., 87
　environment components, 84–89
　global model, 38–45
　immediate value, 47
　infrastructure design, 80–84
　mini, 47
　organizational alignment, 39, 50
　"special purpose," 37, 38
　stepwise construction, 49
　See also Data warehouse environment
Data warehousing, 13
　alternative, 76–77
　background, 14–18
　defined, 17–18

　effective solutions, 26
　history of, 14–16
　importance of, 18
　as migration path, 77
　physical implementation of, 40
　principles, 17
　real world, 169–209
　revolutionary nature of, 17
　rule of thumb, 46
　theory, 32
　USCC, *xvi*
　value-chain driven, 12
　See also Data warehouses
dbPROFILE, 249–50
　customer clustering, 249–50
　defined, 249
　See also DecisionWORKS suite
Decision trees, 144
DecisionWORKS suite, 245–50
　dbPROFILE, 249–50
　information, 245–46, 250
　ModelMAX, 246–49
Dendogram, 135
Descriptive statistics, 131–32
　approaches, 133–37
　cluster analysis, 134–35
　defined, 131
　segmentation method, 136
　using, 131–32
　See also Statistical analysis; Statistics
Distributed computing environment
　　(DCE), 200–201, 203
Drucker, Peter, 27
"Dump and run" model, 81–83
　defined, 81
　failure of, 81–83
　illustrated, 82
　See also Building

Efficiency
　operational, 11, 108, 109–10
　optimization, 32–38

Efficient customer response (ECR) model, 6
Electronic commerce, 8
Electronic data interchange (EDI), 6
Engineering, 158
Enterprise business intelligence, 211–37
   agents, 219
   data discovery, 220
   development tools, 219
   distinguishing features, 215–18
   flexibility and, 218
   many users and, 218
   OLAP, 218–19, 220
   query tools, 219
   reporting tools, 219
   Seagate Holos, 212–13
   statistical analysis tools, 219–20
   system effectiveness, 216
   technology components, 219–20
   visualization, 220
   Web-based discovery, 220
Enterprise modeling, 81
Entity relationship, 16–17
ExchangeInfo, 157
ExchangeInfo Plus, 157
Executive Information Systems (EIS), xv–xvi

Finance and accounting, 61
   knowledge management and, 61
   operational efficiency, 109
   See also Telecommunications systems
Flexibility, 218
Flexible Interactive Reporting Environment (FIRE), 231–34
   defined, 231
   implementation, 231
   replacement of, 233
   See also British Telecom
France Telecom
   billing data visualization, 184–85
   CNET, 170

   competitive environment data visualization, 187
   computed aided telephone interviewing (CATI), 179
   consumer quality of service data visualization, 178–79
   corporate information system of, 171–72
   data reporting, 185–87
   marketing data visualization, 181–83
   modeling visualization, 183–84
   quality of service data visualization, 172–78
   real-time data access/visualization, 187
   responses to open-ended questions visualization, 179–81
   textual data, 181
Fraud/CPS, 196–200
   components, 197
   defined, 197
   final step, 200
   first step, 199
   initiation of, 198
   payback types, 197
Fraud detection, 196–200

Gains chart, 149–51
   cuttoff points, 150
   illustrated, 150
   purpose of, 149
   See also Neural networks
Gains table, 147–49
   creating, 147–48
   illustrated, 149
   interpreting, 148–49
   purpose of, 149
   See also Neural networks
Genetic algorithms, 144
Geographical information systems (GIS), 155
Geographical models, 68
   value chain development for, 69
   See also Business units

GEOlab, 183
   defined, 205
   maps and objects, 205
   See also STATlab
Global model, 38–45
   "architectural templates," 40–41
   checkpoints, 38–40
   developing, 40–42
   development strategy, 45–50
   environment, 48
   illustrated, 39
   logical approach, 41–42
   organizational-based approach, 42
   process-based approach, 42
   stepwise construction, 49
   system-based approach, 42
   value chain analysis and, 42–44
Global positioning system (GPS), 158
Graphical reporting systems (GRS), 155

Histograms, 136
Holistic Systems, 212–13, 225
Holt-Winters method, 184

Inferential statistics, 131, 132–33
   approaches, 137–40
   defined, 132
   regression analysis, 137–40
   See also Statistical analysis; Statistics
Information Junction (IJ), 234
Information superhighway, 8
Information systems, 31
   "architectural templates," 40–41
   France Telecom, 171–72
   value chain alignment with, 71–72
Infrastructure
   design, 80–84
   network, 55–57
   operational, 84, 89
   physical, 84, 89
Interactive and Reporting Information System (IRIS), 230, 232
Internet, 8

Joint application development (JAD), 17
"Kingpin" systems, 72–74
   defined, 72
   illustrated, 73
Knowledge discovery, 96, 97–98
Knowledge management, 6–7, 25–50
   approach, 105–6
   defined, 6–7
   discipline, 23
   framework, 27
   organizational footprint, 29–31
   pioneer, 27
   principles, 27–29
   revolution, 26–31
Knowledge roadmap
   generalized, 53
   solution, 52

LATAInfo, 156
LECInfo, 157
LIDB interfaces, 57
Local exchange carriers (LECs), 3
Logical modeling, 41–42
Long-distance carriers (LDCs), 3

MapInfo Professional, 155–68
   AreaCodeInfo, 157
   CableInfo, 157
   cellsite analysis with, 159–62
   CellularInfo, 157
   ContourInfo, 157
   database query and search tool, 167–68
   defined, 156
   ExchangeInfo, 157
   ExchangeInfo Plus, 157
   exchange location, 166
   fiber analysis, 165–67
   geographical area display, 159
   introduction to, 156–58
   LATAInfo, 156–57
   LECInfo, 157
   Local Market Mapper, 164–65

Long Distance Analysis system, 163
MapMarker, 157
market analysis capabilities, 162–64
MSA/RSAInfo, 157
PCSInfo, 157
StreetWorks, 156
Telecom Mapper, 162
telecommunication offerings, 156–58
TrafficVolumes, 157
Trendline-GIS, 157
views, 163–64
MapMarker, 157
Marketing, 59
  analysis, 158
  data visualization, 181–83
  as driving force, 11
  neural networks and, 145
  programs, 151–52
  role of, 59
  value propositions, 99
  See also Telecommunications systems
Medium-sized business
  business area responsibilities, 64
  organization structure, 64–66
  org chart, 65
Microsoft Excel, 172
  data export from STATlab to, 175
  tracking sales with, 111–14
ModelMAX, 146, 151–52, 246–49
  advanced technologies, 247–48
  defined, 246
  Direct Data Xchange (DDX)
    feature, 248
  functions, 246–47
  learning curve of, 248
  power of, 247
  predictive models, 247, 248
  See also DecisionWORKS suite
MSA/RSAInfo, 157
MS Query, 111–14
  invoking, 112
  query building screen, 112, 113
  report illustration, 114

use example, 112–14
uses, 111
Multidimensional OLAP (MOLAP), ???
Multidimensional scaling (MDS), 202

Network infrastructure
  maintenance, 56–57
  planning and development, 55–56
  value propositions, 101–3
  See also Telecommunications systems
Neural networks, 143–53
  applying, 152
  benefits of, 144
  conclusion on, 152–53
  data input screen, 146
  gains chart, 149–51
  gains table, 147–49
  marketing and, 145
  ModelMAX, 146
  Seagate Holos and, 227–28
  step-by-step use of, 145–52
  training report example, 146–47
  types of, 144
Normalization modeling
    techniques, 16–17

One phone number, 8
Online analytical processing
    (OLAP), 19, 88, 96, 203, 214,
    220–24
  compliance rules, 220
  engine, 222
  enterprise business intelligence, 220
  multidimensional (MOLAP), 222–24
Operational efficiency, 11
  applications, 109–10
  defined, 108
  objective of, 109
  overview, 109–10
Operational infrastructure, 89
  defined, 84
  function of, 89
Operational monitoring, 96–97
  advanced, 117–27

Operational monitoring (continued)
    complex business
        organizations, 118–21
    data preparation and, 121
    data sources, 97
    question formats, 96–97
    reporting levels and, 119–21
Operational systems
    hybridized, 74–76
    as knowledge stores, 76
    resimplifying, 77
Operations, 62
    business, 62
    network, 62
    value propositions, 104–5
    See also Telecommunications systems
Opportunity analysis, 158
Organizational-based model, 42
Organizational structure
    large telecommunications firm, 66
    medium-sized cellular firm, 64–66
    value chain alignment with, 68–71
    value chain and, 63–66

Paging services, 4
Paperless online service tracking
    (POST), 235–37
PCSInfo, 157
Peg counts, 9
Personal communication services (PCS), 4
Phone system
    creating, 55–56
    maintaining, 56–57
Physical infrastructure, 89
    defined, 84
    optimum, 89
PowerCubes, 121
PowerPlay, 117–27
    alerts, 125–26
    defined, 118
    dimension displays, 120
    features, 124–27
    high-level report, 121, 122

    interface use, 124
    performance tracking, 121–24
    PowerCubes, 121
    reporting levels, 119–21
    scheduler, 126–27
Preventative CPS, 180
Private branch exchange (PBX), 3
Proactive CPS, 190–91
Process-based model, 42
Product
    management, 117–27
    performance tracking, 121–24
    See also Sales
Provisioning, 57
    operational efficiency, 109
    value propositions, 104
    See also Telecommunications systems

Quality of service data, 172–78
    consumer, 178–79
    types of, 178
Query tools, 19, 88, 96
    defined, 19
    enterprise business intelligence, 219
    failure of, 115
    MapInfo Professional, 167–68

Regression analysis, 132, 137–40
    assumptions, 140
    coefficients, 139
    defined, 137
    in forecasting, 140
    residual associated with, 184
    results, 138
    running, 137, 138
    See also Statistical analysis
Reporting tools, 19, 88, 96
    defined, 19
    enterprise business intelligence, 219

Sales, 61
    analysis, 107–16, 158
    management, 117–27
    monitoring and control, 110–11

operational efficiency, 109
performance tracking, 101, 121–24
tracking with Microsoft Query/
    Excel, 111–14
value propositions, 101
See also Telecommunications systems
Santa Clause Syndrome, 81
Schedulers (PowerPlay), 126–27
SCRIBE-Questions, 208
SCRIBE-Tutor, 208
Seagate Holos, 212–13
    Ameritech case study, 235–37
    architecture, 224
    British Telecom case study, 229–35
    business intelligence software, 223
    case studies, 229–37
    client engine, 224
    customers, 229
    data mining capability, 228
    defined, 225
    Editor, 232
    functional OLAP products, 221
    hybrid approach, 224
    multidimensional inquiry in, 227
    neural net implementation, 227–28
    RDBMS coupling, 226
    server availability, 228
    SQL links, 228
    two/three-tier applications, 226
    Version 5.0, 227
    See also Online analytical processing
        (OLAP)
Service order processing
    defined, 58
    value propositions, 104
    See also Telecommunications systems
Service rating/ranking systems, 103
Silos
    defined, 30
    illustrated, 31
    optimizing, 32–38
SLP InfoWare Inc., 170, 172
    CNET partnership, 170

Customer Profile System (CPS), 189
Specialized common carriers (SCCs), 3
SPEEDY, 187, 188
SPSS, 239–44
    access and, 240
    action and, 241
    analysis and, 240
    analysts, 241
    automation and, 241
    cluster analysis in, 134
    company information, 241
    as comprehensive solution, 240–41
    consulting services, 241
    dendogram in, 135
    histogram in, 136
    linear regression in, 138
    numerical statistics, 136–37
    offices, 242–44
    regression results, 138
Star Schema design paradigm, 17
Stateside Bell, 69–71
    organizational alignment, 71
    value chain, 70
Statistical analysis
    conclusions on, 140–41
    enterprise business intelligence
        and, 219–20
    large problems and, 143–44
    options and objectives, 131–33
    tools, 19, 88, 97–98
Statistics
    data mining and, 239
    descriptive, 131–32, 133–37
    importance of, 239–40
    inferential, 131, 132–33, 137–40
STATlab
    billing data visualization, 184–85
    boxplots, 176
    business manager availability, 183
    capabilities, 202
    competitive environment data
        visualization, 187

STATlab (continued)
  consumer quality of service data
      and, 178–79
  CSQlab, 206–7
  as data analyzer, 174, 201–2
  data availability and, 200–201
  DATAboard, 178, 207–8
  for data exploitation, 179, 180
  data export from, 175
  DATAman, 203
  DATAread, 203–5
  for data reporting, 179, 180
  data reporting, 185–87
  Data Warehouse Solution, 170–71
  enduser tools, 201–8
  GEOlab, 205
  information object architecture, 201
  macro language, 202
  marketing data visualization, 181–83
  modeling visualization, 183–84
  quality of service data
      visualization, 172–78
  real-time data access/
      visualization, 187
  responses to open-ended questions
      and, 179–81
  SCRIBE-Questions, 208
  SCRIBE-Tutor, 208
  spreadsheet, 175
  spreadsheet data manager, 202, 206
  TIMElab, 205–6
  time series report, 177
  tools, 170–209
Storage area, 84, 87
  characteristics of, 87
  See also Data warehouse environment
StreetWorks, 156
Structured query language (SQL), 19
Switching system, 74–76
System-based model, 42
System interdependency
  complexity of, 37
  functioning of, 36

  illustrated, 36
Systems development life cycle, 81

Technical proficiency, 12
Technological excellence, 108
Telecom Mapper, 162
Telecommunications
  analysis dependency, 21
  business of, 2
  competitive climate of, 21
  data intensity, 20
  dependence on, 5
  as driving economic force, 5–6
  "experiments," 7–8
  future directions, 8
  geographical information and, 158–59
  impact of, 2
  industry segments, 3–4
  technological change and, 21
  technological innovation and, 9
  value chain, 51–78
  value propositions in, 95–106
Telecommunications systems, 53–63
  acquisition, 54–55
  activation, 57–58
  billing, 58–59
  creation, 54
  credit management, 61–62
  customer service, 59–61
  finance/accounting, 61
  marketing, 59
  network infrastructure
      maintenance, 56–57
  network infrastructure planning/
      development, 55–56
  operations, 62
  provisioning, 57
  sales, 61
  service order processing, 58
TIMElab, 184, 205–6
  defined, 205
  functions, 206
  See also STATlab

Traffic engineering, 102–3
TrafficVolumes, 157
Trendline-GIS, 157
United States Cellular Corporation (USCC), *xvi*
Value added carriers (VACs), 3
Value chain, 51–78
   analysis, 42–44
   business unit allocation to, 66–78
   comprehensive, 62–63
   example, 44
   for geographically based business unit, 69
   illustrated, 63
   information system alignment, 71–72
   organizational alignment, 68–71
   organizational structure and, 63–66
   process steps, 52–53
   Stateside Bell, 70
   telecommunications functions/systems and, 53–63
   tracing back to customer, 69–70
   walking through, 44–45
Value propositions
   activation, 104
   billing, 104
   by functional area, 98–105
   creation, 103–4
   credit, 99–100
   customer service, 100
   defined, 90
   gathering, 90–92
   knowledge discovery, 97–98
   marketing, 99
   network maintenance, 103
   network planning, 101–3
   operations, 96–97, 104–5
   provisioning, 104
   sales, 101
   service order processing, 104
   in telecommunications, 95–106
Visualization, 19, 88, 98
   of billing data, 184–85
   of consumer quality of service data, 178–79
   of enterprise business intelligence, 220
   of marketing data, 181–83
   of modeling, 183–84
   of quality of service data, 172–78
   of real-time data, 187
   of responses to open-ended questions, 179–81
Web-based discovery, 19, 88, 97
   defined, 19
   enterprise business intelligence, 220
Wide area telephone services (WATS), 3
World Wide Web (WWW), 8

The Artech House Computer Science Library

Applied Maple for Engineers and Scientists, Chris Tocci and Steve Adams

ATM Switching Systems, Thomas M. Chen and Stephen S. Liu

Authentication Systems for Secure Networks, Rolf Oppliger

Client/Server Computing: Architecture, Applications, and Distributed Sytems Management, Bruce Elbert and Bobby Martyna

Computer-Mediated Communications: Multimedia Applications, Rob Walters

Computer Telephone Integration, Rob Walters

Data Modeling and Design for Today's Architectures, Angelo Bobak

Data Quality for the Information Age, Thomas C. Redman

Data Warehousing and Data Mining for Telecommunications, Rob Mattison

Distributed and Multi-Database Systems, Angelo R. Bobak

Electronic Payment Systems, Donal O'Mahony, Michael Peirce, Hitesh Tewari

A Guide to Programming Languages: Overview and Comparison, Ruknet Cezzar

Heterogeneous Computing, Mary M. Eshagian, editor

Internet and Intranet Security, Rolf Oppliger

Internet Digital Libraries: The International Dimension, Jack Kessler

Introduction to Document Image Processing Techniques, Ronald G. Matteson

Managing Computer Networks: A Case-Based Reasoning Approach, Lundy Lewis

Networks and Imaging Systems in a Windowed Environment, Marc R. D'Alleyrand

Practical Guide to Software Quality Management, John W. Horch

Risk Management Processes for Software Engineering Models,
  Marian Myerson

Software Verification and Validation: A Practitioner's Guide,
  Steven R. Rakitin

Successful C for Commercial UNIX Developers, Mohamed Osman

Survival in the Software Jungle, Mark Norris

UNIX Internetworking, Second Edition, Uday O. Pabrai

Wireless LAN Systems, A. Santamaría and F. J. López-Hernández

Wireless: The Revolution in Personal Telecommunications, Ira Brodsky

X Window System User's Guide, Uday O. Pabrai

For further information on these and other Artech House titles,
including previously considered out-of-print books now available through our
In-Print-Forever™ (IPF™) program, contact:

Artech House
685 Canton Street
Norwood, MA 02062
781-769-9750
Fax: 781-769-6334
Telex: 951-659
email: artech@artech-house.com

Artech House
Portland House, Stag Place
London SW1E 5XA England
+44 (0) 171-973-8077
Fax: +44 (0)171-630-0166
Telex: 951-659
email: artech-uk@artech-house.com

Find us on the World Wide Web at:
www.artech-house.com